Angelo Artale

Rings in Auctions

An Experimental Approach

Springer

Author

Dr. Angelo Artale
University of Bonn
Wirtschaftspolitische Abteilung
Adenauerallee 24–42
D-53113 Bonn, Germany

Library of Congress Cataloging-in-Publication Data

Artale,Angelo, 1969-
    Rings in auctions : an experimental approach / Angelo Artale.
        p.   cm. -- (Lecture notes in economics and mathematical
    systems ; 447)
    Includes bibliographical references.
    ISBN 978-3-540-61930-7   ISBN 978-3-642-59158-7 (eBook)
    DOI 10.1007/978-3-642-59158-7
    1. Auctions--Mathematical models. 2. Economics, Mathematical.
    3. Game theory.   I. Title. II. Series.
    HF5476.A77   1997
    330'.01'5193--dc21   96-46237
                                                              CIP

ISSN 0075-8442
ISBN 978-3-540-61930-7

© Springer-Verlag Berlin Heidelberg 1997
Originally published by Springer-Verlag Berlin Heidelberg New York in 1997

The use of general descriptive names, registered names, trademarks, etc. in this publication does not imply, even in the absence of a specific statement, that such names are exempt from the relevant protective laws and regulations and therefore free for general use.

Typesetting: Camera ready by author

SPIN: 10546692   42/3142-543210 - Printed on acid-free paper

Ai miei Genitori

Lecture Notes in Economics and Mathematical Systems 447

Springer-Verlag Berlin Heidelberg GmbH

Acknowledgments

I am very in debt to Reinhard Selten for having stimulated my interest in Experimental Economics, for his comments and helpful discussions. I would like to thank Urs Schweizer for his encouragement. I thank also all my collegues of the *Graduiertenkolleg*, European Doctoral Program and Laboratory for Experimental Economics at the University of Bonn. For helpful comments, I am in debt to Torkild Byg, Vincent Crawford, Peter Funk, John Kagel, Ulrich Kamecke, George Mailath, Abdolkarim Sadrieh, Karl Schlag and Heide Will. Gary Bolton, Alicia García Herrero, Bettina Kuon and Heide Will helped me to improve the draft manuscript. Many useful suggestions were offered by the participants of the workshops at Gießen (Rauischholzhausen) and Stony Brook in July 1995. Financial support from the *Deutsche Forschungsgemeinschaft* is gratefully acknowledged. Finally, I wish to thank Ulrike Haupt und Petra Moos who, with great patience, corrected the protocols.

Contents

Chapter 1

Introduction

In auctions, bidders compete with one another in their attempt to purchase the goods that are up for sale[1]. But buyer competition may be reduced or disappear when a ring of colluding bidders is present. The purpose of the participants to a ring is to eliminate buyer competition and to realize a gain over vendors. When all participants are members of the ring, this is done by purchasing the item at the reserve price and splitting the spoils (the difference between the item market value and the reserve price) among the participants. "The term *ring* apparently derives from the fact that in a settlement sale following the auction, members of the collusive arrangement form a circle or ring to facilitate observation of their trading behavior by the ring leader" (Cassady jr. (1967)). If the coalition members knew other players' values, the problem faced by the ring might be easily solved: the player with the highest value should submit a serious bid and the other members, on the contrary, only phony bids. However, ring participants do not usually know the values of other members. Therefore, ring members have to find out some mechanism which selects the player who has to bid seriously and, eventually, establish side payments paid to each of the losers[2].

Since we are interested in determining the outcome (the selected winner, his bid and, eventually, the side payments) of a given mechanism, we also need a description of how individuals interact under the specific rules of a mechanism. The approach taken in the Bayesian implementation (Palfrey and Srivastava (1993)) is based on Bayesian game theory

---

[1]Surveys of the auctions literature are found in McAfee and McMillan (1987) and Wilson (1992).

[2]We are implicitly assuming that all bidders are also ring members. When it is not the case the problem becomes more complicated: the mechanism has to select the serious bidder, establish his bid, and, eventually, side payments paid to each of the losers if the ring gets the item.

(Harsanyi (1967-68)). In this approach, the preferences and information of agents are modeled explicitly, beliefs are updated according to the Bayes' rule, outcome are evaluated according to expected utility theory, and finally, actions are chosen to maximize expected utility given the action choices of other agents. Equilibrium is defined as a collection of strategies in which each agent is choosing a best strategy given the strategies of the other agents.

The problem of characterizing all mechanisms which implement a given outcome would seem formidable, since the set of possible mechanism is extremely large. Moreover, the problem is complicated by the fact that the mechanism itself is the choice variable. These problems have been overcome by the theory of mechanism design by the so-called revelation principle which reduces the number of mechanisms to a small and specific class. The revelation principle states that the outcome of any mechanism that is not incentive-compatible can be mimicked by one that is incentive-compatible, so that honesty can be assumed without loss of generality. A direct mechanism, then, takes the vector of bidders' reports of values and dictates bids and (perhaps) side payments to each bidder. Unfortunately, direct mechanisms may have many equilibria, some of which give undesirable outcomes. The most recent work on Bayesian implementation tries to overcome this problem enlarging the space of messages that each player may send (Palfrey (1992)).

Summing up, the theory of Bayesian implementation assumes that players update their beliefs according to Bayes' rule, that they maximize their expected utility and are Nash competitors. Moreover, in virtue of the revelation principle, it restricts its focus on (incentive-compatible) direct mechanisms.

Numerous experimental results in the last decades have shown that human behavior violates the assumption of Bayesian game and decision theory[3]. The fact that all the assumptions of the Bayesian theory of implementation are usually violated seems to be a reason to believe that we need a descriptive theory of implementation. As we have already seen, the aim of implementation theory is to find a game (mechanism) which gives as (possibly unique) equilibrium outcome the desired allocation. In this sense, a descriptive theory of implementation gives an answer to the following question:"Which mechanisms, if any, are likely to be observed when human subjects have to solve a given implementation problem?". Until today, one of the most important task of Experimental Economics has been to reveal whether human behavior agrees with game theoretical

---

[3]To my knowledge, the most clear synthesis of different position in the literature about economic behavior can be found in Selten (1991).

prediction within a given framework, that is, whether players' action, given the rule of the game, conform with the game theoretical prediction. To our knowledge, our approach is innovatory. It allows to investigate not only beliefs updating and strategy choice but also the discovery of game mechanisms which players spontaneously construct. This approach allows us to find out whether there are social rules (mechanisms) which have a higher degree of social acceptance than others. That is, we can answer to the question whether there are mechanisms which are more frequently used than others.

The implementation problem we consider is that faced by a ring of three players who participate in a first-price sealed-bid auction. Players' values are and remain private information. After all players have learned their values but before they submit their bids they are allowed to communicate as long as they want. They may not sign contracts. Players learn only who is the winner and his bid. The winner is allowed to make side payments. The seller's behavior is assumed to be passive. He announces a reserve price r before the auction is played. We want to find out, in this way, which mechanisms, if any, do experimental subjects use when they are allowed to collude and to make side payments. In each round and for each group, we assure that at least one experimental subject has a private valuation equal to or higher than r, and it is common knowledge. This feature of the experimental design has two effects. It assures that ex ante all groups may cooperate for the same number of rounds but, as we will see, it can complicate the analysis of some mechanisms when players do not simultaneously choose their strategies.

Rings' building in first-price auction has been theoretically investigated by McAfee and McMillan (1992) when all bidders are coalition members. They characterize the optimal direct mechanism, that is, the direct mechanism which selects as a winner the player with the highest value. This mechanism can be mimicked by an indirect mechanism. Before players play the legitimate auction, they play a prior first-price auction. If the highest bid in this prior auction exceeds the reserve price, the winner then bids the reserve price in the legitimate auction and pays each of the losers an equal share of the difference between his bid in the prior auction and the reserve price.

We will see that the mechanism that experimental subjects used most frequently is a non-incentive-compatible and non-optimal mechanism. Nevertheless, since experimental subjects announce their true values in most of the cases, they reach an almost optimal result.

We have organized the material in the following way: Chapter 2 introduces the experimental design and the experimental results. We describe the mechanisms used by experimental subjects and compare experimental data with game theoretical predictions. A descriptive model of the most used mechanism is presented in Chapter 3. Chapter 4 analyzes the observed mechanisms from a strategic point of view. We assume that, given that experimental subjects have chosen a mechanism, they will play according to it. We also address the problem of the optimality of the chosen mechanisms. We rank the observed mechanisms according to their ex ante expected payoff. Chapter 5 considers two extensions of the mechanisms used by experimental subjects. Finally, Chapter 6 concludes summing up the main results and comparing our work with the most recent empirical work on collusion in auction and with experimental work on designer markets.

Chapter 2

The Experiment

This chapter is organized as follows. Section 2.1 sums up the most recent contributions on collusion in auctions. The experimental set-up and technology are described in Section 2.2. Section 2.3 reports the theory of collusion in first-price auctions with emphasis on the contribution of McAfee and McMillan (1992). Section 2.4 describes the mechanisms used by experimental subjects. That is, we describe the rules chosen by experimental subjects to implement collusion. Section 2.5 strategically analyzes the mechanisms used by experimental subjects. We assume that, given that players have chosen some mechanism, they will play according to it. We ask which is the best strategy for the proposed mechanism. We compare the experimental data with the theoretical predictions. For each observed mechanism, we calculate the ex-ante expected payoff when players play according to the theoretical prediction, and compare this with the observed payoff on average. As we will see, the most commonly used mechanism is not an optimal one, that is, it is not a mechanism which selects as winner the player with the highest value when players play rationally. Nevertheless, because of the bounded rationality of experimental subjects, it reaches the efficient result in most of the cases.

Besides these mechanisms, which have been used by experimental subjects, there are other mechanisms that were proposed in the cheap talk phase but not used. In Section 2.6, we describe these other mechanisms and try to explain the reason of their failure. Finally, Section 2.7 analyzes the relation between experimental subjects' strategic behavior and the so-called "end effect".

2.1   Introduction and Related Literature

Since the seminal work of Vickrey (1961), there has been extensive development of the theory of non-cooperative auctions[1]. In recent years, there has also been an increasing amount of empirical testing of the theory of non-cooperative auctions[2]. Only during the last few years, scientific work appeared which tried to understand bidders' and seller's behavior from a normative point of view, when the bidders collude.

In one of the first theoretical contributes on collusion in auction, Robinson (1985) argues for the relative non-susceptibility of first-price auctions to collusion by bidders. Collusion in English auction is analyzed by Graham and Marshall (1987) who propose the following collusive mechanism.

> "Each of the l ring members receives a fixed non-contingent payment P from the center. They each then report a number to the center. The center instructs the ring member with the highest report to remain active up to that report at the main auction. All other ring bidders are instructed to not bid or bid zero. If the coalition wins, then the winner pays the center the maximum of zero and the difference between the second highest pre-auction report of ring members and the price paid at the auction. If the coalition does not win, no payment is made to the center."

This mechanism is (individually rational and) incentive compatible, that is, each ring member reports his true value to the center. Although this mechanism is ex ante budget balanced (feasible), it needs a "center" which also plays the role of a potential ex post budget breaker. In their work, Graham and Marshall assume that all the bidders are homogeneous. This hypothesis is removed by Mailath and Zamsky (1991) who characterize collusion in second price auctions when the bidders are heterogeneous. They characterize a mechanism that is ex post feasible. All these papers analyze collusion in auction in the simple case that a single indivisible item has to be sold. Collusion in multiple object uniform auctions[3] is empirically investigated by Baldwin et al. (1995).

---

[1] Milgrom and Weber (1982a) provide a very general framework for analyzing non-cooperative auctions and compare different auction forms and different seller policies.

[2] See above all Kagel's survey (1995).

[3] In an uniform auction the k highest bidders pay all the $(k+1)st$ bid. For multiple object auctions, see also Milgrom and Weber (1982b) and Weber (1983).

Collusion in first-price auctions has been analyzed by McAfee and McMillan (1992). They have characterized (in a theoretical work) coordinated bidding strategies in a first-price auction setting[4]. As we will see in Section 2.3, this mechanism is optimal, that is, it selects always as winner the player with the highest value. Güth and Peleg (1993) proceed with an axiomatic approach; they show, using the axiom of "envyfreeness with respect to bids", that collusion in first-price auctions is not implementable. Collusion in first-price auction with a special kind of heterogeneity is considered by Marshall et al. (1994). They analyze numerically collusion in first-price auction in the case in which one coalition with k_1 participants plays versus another coalition with k_2 participants[5] (with $k_1 \neq k_2$), and in the case in which a coalition with k_1 participants plays versus the remaining part of players (and each of them is assumed to act non-cooperatively). They assume that players' values are independently drawn from a uniform distribution on $[0, 1]$ and therefore analyze the problems as two different particular cases of asymmetric first-price auctions.

The meager number of experimental work in which collusion is allowed can be explained by a number of significant obstacles: communication among bidders is typically not allowed, and subjects are usually brought together for a single auction session, or in case of repeated sessions, the group composition typically changes between sessions.

Isaac and Walker (1985) try to induce implicit collusion on experimental subjects. Subjects play a discriminative price sealed bid auction and are allowed to communicate. Nevertheless there are some restrictions. Subjects are free to discuss any aspect of the experiment they wish, except that:

1. they may not discuss any quantitative aspects of the private information on their screen.

2. they are not allowed to discuss side payments or physical threats.

Our work is about an experimental investigation of collusion in the first-price sealed bid auction setting with private values when unexperienced experimental subjects are *explicitly* allowed to collude and to make side payments. We want to find out which mechanisms are used by experimental subjects in such a setting. Theoretical possibilities have not been

---

[4]They characterize coordinated bidding strategies also in the case in which the bidders cannot make side payments.

[5]They assume that the values of each player are common knowledge within a coalition, and hence they do not have to solve any implementation problem.

explained to them. Subjects have to invent the mechanism they want to use themselves. They are expected to build coalitions. Each coalition has to solve the following problem. A potential winner of the auction and a rule for dividing the spoils are to be chosen. To do it, the coalition needs a mechanism which selects a bidder, (possibly the one with the highest valuation) and which promises to each player an expected payoff at least equal to a certain disagreement value. To our knowledge, this experiment is the first one that partly endogenises the rules of the game, giving the players the possibility to design them.

2.2   Experimental Design

This experiment was run at the end of 1994 (15 sessions) and in October 1995 (1 session) in the laboratory of the University of Bonn. Sixteen groups, each with 3 players, participated in a first-price sealed bid auction. Participants were students of the University of Bonn. Table 2.1 shows the students' distribution according to their academic background. Upon arrival, subjects were given copies of the instructions for the session (see Appendix A). The instructions were read aloud and questions answered in public to make all instructional information common knowledge. Each group received the same following information about the auction.

INSTRUCTIONS

- A single item was to be sold. Each participant might make only one bid.

- The player who made the greatest bid was the winner and got a payoff equal to his valuation minus his bid. If more players made the highest bid, a fair dice was thrown to choose the winner. Monetary values for the item being auctioned were induced on subjects

Years of Study	*Economics*	*Law*	*Math & Sciences*	*Other*
1-2	8	14	2	0
3-5	17	4	2	1
Game Theory (Yes)	12	0	0	0

Table 2.1: Academic background of experimental subjects

using the concept of "resale values". That is, each subject was told that for any winning bid the bidder would receive a profit equal to his resale value minus his bid.

- Each player knew his valuation only. Players' valuations were drawn from a joint uniform discrete distribution on $\{51, .., 100\}$. Subjects learned their own valuations drawing one card, which had to be returned soon afterwards, from a pack of 50 playing-cards. Players were not allowed to show their private valuations to other players. The only restriction to players' bids was that the bid of each player, b_i, had to be equal to or lower than 100. Values remained private information after the end of the session too.

- After they had learned their private valuations, subjects were allowed to communicate as long as they wanted. They might not sign contracts. While subjects communicated, they were filmed.

- After that players made their bids, the experimenter announced who was the winner and his bid.

- Afterwards, the winner might make side payments to losing bidders. Total side payments might at most be equal to the realized payoff in each round. The experimenter announced in each round the amount of money that was paid from the winner to each losing bidder.

- The same auction was played 20 times by the same group of players. At the beginning of each round, new valuations were drawn, independently of the valuations of the previous rounds.

- The seller's behavior was assumed to be passive. He announced a reserve price r equal to 58 before the auction was played. In each round and for each group, we assured that at least one experimental subject i had a private valuation $v_i \geq 58$, and it was common knowledge. The reserve price remained unchanged in each round.

- Gains were calculated in *soldi*, (1 soldo=20 Pfennig). The exchange rate was calibrated in order to ensure on average 40 DM to each of players, if they cooperated optimally, that is, if the winner always bid r. Subjects were also endowed with 10 *soldi* to compensate any loss. Altogether, players were not allowed to lose more than 10 *soldi*[6].

---

[6]Old European coin (golden *soldo*) in use among Goth, Frank and Lombard people. In Carolingian age, it was equivalent to one twentieth of one *lira*.

TECHNOLOGY

- After each player drew his value, this was saved in a file. We guaranteed that $v_i \geq 58$ for at least one player, repeating the round if all three values were lower than the reserve price.

- Subjects were filmed with a videocamera. To identify players in the protocols, we named the player who sat on the right side in front of the camera Player 1, the players who sat in the middle and on the left Player 2 and Player 3, respectively (see Artale (1996)).

- Bids and side payments were made by experimental subjects through forms which have been reproduced in Appendix A. Making their bids, players were invited to motivate shortly their decision.

- In all experiments, subjects were inexperienced. That is, they had not previously participated in a laboratory experiment using sealed bid auctions. In recruiting people, we have been careful to ensure that players in each group had different academic backgrounds and, to the best of our knowledge[7], that they did not know each other.

- Experimental sessions lasted approximately two to three hours. The instruction time was 10 to 15 minutes.

Some remarks about the experimental design follow. We assume that the auctioneer announces a positive reserve price as it is usually assumed in the auctions literature. Note that the probability that all three players have a value which is lower than r is equal to $(7/50)^3$. Assuring that at least one player in each round has a value which is equal to or higher than the reserve price, we lose the independence of the values but we gain the certainty that ex ante all groups may cooperate for the same number of rounds. However, the loss of the independence of the values will complicate the analysis of some mechanisms in which the players do not simultaneously choose their strategies. To clarify this point, we anticipate something that we will analyze in detail in Chapter 4. Consider the following mechanism which has been really used by experimental subjects. Players announce their values in sequence. The player who announces the highest value pays the reserve price in the legitimate auction and splits equally the difference between his announced value and the reserve price into three parts and gives one part other players. Now, if the

---

[7]Participants presented themselves to the experiment registration. They were not informed about other participants before the game began.

player who announces second has drawn a value equal to, say, 54 and the player who announces first has announced 51, then to calculate his best announcement, the player who announces second has to know whether the player who announces first has a value lower than 58. In fact, if the player who announces second knows that the player who announces first has a value lower than 58, then he can conclude that the player who announces third has a value equal to or higher than 58. If the values were independent, this would not happen. The distribution function of the values of the player who announces third would not depend on the values drawn by the two other players.

One might argue that our experimental design does not make an experiment on mechanism design possible. In fact, one could argue that players should be allowed to discuss ex ante, i.e. when they do not yet know their values. They should choose the mechanism according to what they want to play, draw the values and have again the possibility to communicate in order to implement the mechanism. There are different reasons which have induced us to choose this particular experimental design. The most compelling reason is that, since the game is repeated, it does not matter (for the last nineteen rounds) whether experimental subjects are allowed to discuss before they know their values. The second reason is that allowing subjects to discuss before values are learned is useless if we do not allow them to sign contracts; subjects could decide to use another mechanism after they have learned their values[8]. Since we do not consider it very realistic to allow players to sign contracts for such agreements, we have opted for our design. The third reason is that our design is relatively easy to understand. Easiness of understanding is an important experimental feature. The better experimental instructions can be understood by experimental subjects, the more confidence we can have in experimental data.

As we have seen above, side payments may at most be equal to the realized payoff in each round. This restriction excludes ex ante some mechanisms (which are also used in the real world—see Cassady jr. (1967)) for instance where players choose the winner using a random device (like the face of the moon or a dice) and the winner has to pay a fixed amount of money to each of the losers. We are aware of the fact that it would have been more realistic to allow experimental subjects to make some losses in making side payments, but it would have made our game quite

---

[8]Actually, allowing subjects to discuss twice without binding agreements could make the evaluation of experimental data more difficult. In fact, proposing to switch from one mechanism to another could be interpreted as a signal that a player has drawn a good or a bad value.

difficult for theoretical analysis. In fact, allowing losses in making side payments could make room for dynamic strategies.

The Australian wool trade market is a real situation of collusion which could be in some sense similar to our experiment. There appears to be a strong tendency among buyers of wool to organize rings, or so-called *pies*. There are hundreds of types of wool, and the large total number of buyers in any one market shrinks to only a few when related to the purchase of a particular type and grade. Usually, buyers make arrangements with a small number of buyers and belong to a number of pies, each having a different combination of buyers (Cassady jr. (1967)). *Mutatis mutandis*, one can imagine that our players engage in each round in the purchase of a particular type of wool, but they do not know their values for the next round until it starts.

As we have seen in Section 2.1, Isaac and Walker (1985) tried to induce some kind of implicit collusion in their work. In our design, in contrast, we tried to induce explicit collusion.

2.3   Theory of Collusion in First-price Auctions

We now want to look in a formal way at collusion in auctions when the values are private information and side payments are allowed. A very general model should consider the implementation problem that each coalition has to solve, the strategic interactions inside and outside the coalition, and the seller's response.

In order to define an implementation mechanism, we have to introduce some notation. Let $N = \{1, \ldots, n\}$ be the set of players and T_i be the set of types of Player i. Let A be the set of feasible alternatives, X the set of allocation rules $x : T_C \to A$ where $T_C = \times_{i=1}^{C} T_i$ and $C \subseteq N$ is the set of players who participate in some coalition. Let $\mu = (M, g)$ be a mechanism, where $M = M^1 \times \cdots \times M^C$, and g is a function $g : M \to A$. The set M^i is the *message space* of Player i, and g is called *outcome function*. For each $m \in M$, $g(m)$ yields an outcome in the set of alternatives. Solving the implementation problem consists of finding some mechanism (M, g) and some Bayesian equilibrium σ of (M, g) such that $g(\sigma) = x^*$ [9]. In our problem $M^i = V_i = V$, for all $i \in C$, where V_i is the set of values of Player i, and $x = (\rho, \xi)$, with $\rho : V_C \to [0, 1]^C$ and $\xi : V_C \to \Re^C$, where $V_C = \times_{i=1}^{C} V_i$. Let $\hat{v} = (\hat{v}_1, \ldots, \hat{v}_C)$ be the vector of

---

[9]We will say in this case that the allocation rule x^* is *weakly implementable via Bayesian equilibrium* (see Palfrey and Srivastava (1993)).

messages sent by the coalition members. When Player i sends a possibly untrue report $\hat{v}_i \in V$ of his valuation v_i to the mechanism, $\rho_i(\hat{v})$ is the probability that Player i is the sole bidder of the coalition C and $\xi_i(\hat{v})$ is his (expected) received side payment [10]. We require $\sum_{i \in C} \rho_i(\hat{v}) \leq 1$, for all $\hat{v} \in V_C$. By the revelation principle, the restriction to direct revelation mechanisms is without loss of generality: any Nash equilibrium outcome of any game which determines the sole bidder and side payments will also be a Nash equilibrium outcome of some direct revelation game in which the bidders report their valuations truthfully.

When $C \subset N$, we also have to consider the strategic interaction of the coalition C against other coalitions or single players or both. Modelling this problem as a non-cooperative game, one meets technical difficulties (like systems of differential equations which do not have analytical solutions) which are difficult to overcome. Modelling this problem as a cooperative game, other difficulties can arise. In fact, since the collusive payoff of each coalition depends on the values of the bidders a cooperative game may not be well defined when these are private information. Moreover, since not always subcoalitions form a partition of the set of players (because of nesting and overlapping of coalitions), cooperative game theory cannot solve the problem. Some authors (see Holmström and Myerson (1983)) try to overcome these problems by saying that intelligent players in a cooperative game with incomplete information would themselves bargain over the set of incentive-compatible and durable mechanisms[11]. Thus, a theory of cooperative games with incomplete information should be a theory of mechanism selection by individuals who have private information.

In the literature on mechanism design with incomplete information, it is usually assumed that all agents adopt Bayesian-Nash behavior. As we have seen above, one requires that the desired allocation rule is a Bayesian-Nash equilibrium of the mechanism μ. This behavior requires a great deal of ability on the part of the agents. As far as we know, there are no "non-Bayesian" explorations of implementation with incomplete information. As we pointed out in Section 2.1, this work wants to investigate, which mechanisms, if any, are used by experimental subjects. In

---

[10]We always assume the bidders have homogeneous values (i.e. the values are drawn all from the same distribution) not because this assumption seems to us to be realistic but because the implementation of collusion with heterogeneous bidders is more complex and, to our knowledge, in some cases, it has not yet been solved (see Mailath and Zemsky (1991)).

[11]We say that a mechanism is durable if its outcome is durable, that is, if it resists against an unanimous rejection.

this sense, we can say that we try to find out "natural" (not necessarily direct and/or Bayesian) mechanisms, which are expected to be used by experimental subjects.

Another point refers to the information of ring participants about non-ring participants. Should we assume that non-ring participants are aware of the existence of the ring or not? And if they are aware, should they know the membership of the ring? What about the beliefs of the auctioneer? Should we assume that he knows something about the existence and the size of the cartel? Many of these questions can find an answer or, at least, suggestions in field studies.

In this work, we will not try to examine all these single aspects. As we have seen, we restrict ourselves to the case in which there are only three players who participate in the auction. By having a small number of participants, we try to make the grand coalition very likely to be observed. The behavior of the auctioneer is assumed to be passive. These assumptions and others specified above make the problem very specific. We are aware of this fact and hope for future research to fill these gaps.

McAfee and McMillan (1992) have characterized the optimal direct mechanism in this setting. They suppose that a unique item is to be sold by a sealed bid auction to one of a set of risk-neutral bidders. They define an *optimal* cartel mechanism to be a mechanism with the following property: the bidder with the highest value wins if and only if his value exceeds the reserve price r and the seller receives r. The optimal direct mechanism that implements this is as follows.

Let us assume that the private valuations are independently drawn from the same cumulative distribution function F on $[0, \overline{v}]$, where F has a differentiable density f. Let n be the number of bidders. Then the following theorem holds.

Theorem 1 (McAfee and McMillan (1992)) *The following mechanism is incentive-compatible and efficient. Before the auction, the cartel members report their valuations to the mechanism. If no report exceeds r, the cartel does not bid in the auction. If at least one bid exceeds r, the bidder making the highest report v obtains the item and pays a total of*

$$T(v) = F(v)^{-n} \times \int_r^v (u - r)(n - 1)F(u)^{n-1}f(u)du + r \qquad (2.1)$$

Each losing bidder receives from the winner $[T(v) - r]/(n - 1)$, and the seller receives r.

This mechanism is not the only Bayes-Nash implementable and efficient one. The cartel can set up an auction of its own as the following corollary shows.

Corollary 1 (McAfee and McMillan (1992)) *The following mechanism is also Bayes-Nash implementable and efficient: the players organize a prior sealed bid first-price auction among themselves. If the highest bid in this prior auction exceeds r, the winner then bids r in the legitimate auction and pays each of losers an equal share of the difference between his bid in the prior auction and r, keeping nothing of this difference for himself.*

To prove this corollary, note that, in the new mechanism, bidding $T(v)$ is an equilibrium because, if all others bid $T(v)$, bidding $T(v)$ in the new mechanism has the same effect as responding honestly in the direct mechanism.

It is surprising that efficiency can be achieved in the light of Myerson and Satterthwaite's (1983) general result, that transfers do not guarantee efficiency in general bargaining situations when individual rationality is required.

This analysis is valid only if all the subjects engage in the grand coalition. McAfee and McMillan examine the case also in which only some of them are in the cartel, but they cannot characterize the bid of coalition members for a general distribution function F (see Pesendorfer (1994a) and Pesendorfer (1994b) for a more general case of rings in procurement auctions for school milk in Florida and in Texas).

2.4 Experimental Results

In this section we start to present the experimental results. We will proceed as follows. In Subsection 2.4.1, we ask whether experimental subjects cooperated. We present two coefficients of cooperation. In Subsection 2.4.2, we describe the mechanisms spontaneously designed by experimental subjects. Subsection 2.4.3 reports winners' bids and payoff shares. Subsection 2.4.4 considers the case of a two-player coalition which plays versus one individual bidder. Finally, Subsection 2.4.5 concludes, summing up the most important results.

2.4.1 Do Experimental Subjects Cooperate?

It is the aim of this subsection to establish whether experimental subjects cooperated, that is, whether they agreed on some collusive rule.

Since subjects could not enforce their agreements it is important to distinguish between ex ante and ex post cooperation, i.e., it is important to distinguish between agreements reached which are not implemented and agreements reached which are implemented.

Definition 1 *The* ex ante *coefficient of cooperation, KK, is the relative frequency with which the players agree on some collusive mechanism, before they make their bids.*

Definition 2 *The* ex post *coefficient of cooperation, KK^*, is the relative frequency with which the players agree on some collusive mechanism and implement it.*

It follows from the definition that $KK^* \leq KK$. The average values of these two coefficients are $MKK \simeq 0.95$ for the ex ante coefficient of cooperation and $MKK^* \simeq 0.92$ for the ex post coefficient of cooperation. Table 2.2 reports in detail the value of the two coefficients in each session.

Note that the backward induction solution of this auction as a supergame is trivial: play always non-cooperatively. In fact, since players may not sign any contract, there is no incentive to stick to any agreement in the last round. In the penultimate round, rational players will anticipate the behavior in the last round and will not stick to any agreement. Using the same argument, one can go back until the first round, where there is no incentive to stick to any agreement. Nevertheless, we have observed that experimental subjects cooperate until shortly before the end of the game. Different explanations for such kind of behavior have already been given in the literature (Kreps et al. (1982), Selten

	A_1	A_2	A_3	A_4	A_5	A_6	A_7	A_8
KK	1	1	0.9	1	0.9	0.8	1	1
KK^*	0.9	0.95	0.85	1	0.85	0.8	1	1

	A_9	A_{10}	A_{11}	A_{12}	A_{13}	A_{14}	A_{15}	A_{16}
KK	0.85	1	0.8	1	1	1	0.9	0.95
KK^*	0.85	1	0.65	1	1	1	0.9	0.9

Table 2.2: Ex ante and ex post coefficient of cooperation in sixteen sessions $A_1, \ldots, A_{16}$

and Stoecker (1986), Selten (1990)). Assuming that experimental subjects use a trigger strategy[12], we calculate the ex ante expected round in which cooperation breaks down.

The ex ante expected payoff of each player in each round is roughly equal to 10 *soldi* (see Chapter 4) when players cooperate, that is when they bid the reserve price r and split the difference of the highest value minus r. Each player's non-cooperative ex ante expected payoff in each round is approximately equal to 4,20 *soldi*. Suppose that a player believes that the other players will continue to cooperate, as long as he cooperates and will play non-cooperatively for all remaining periods, if he defects. One can easily calculate that under this assumption, it is optimal to defect in period 17 [13]. Looking only at the auctions in which players deviated, we found that Round 17 was, on average, the first round in which players defected after having reached an agreement. That is, the first round in which cooperation broke down (on average) is equal to the first round in which cooperation would break down if players used a trigger strategy. Looking at all auctions, Round 18 was, on average, the first round in which players deviated after having reached an agreement[14].

In all auctions, but two, experimental subjects cooperated at least for the first 16 rounds, that is, they reached an agreement and implemented it. Auction 6 was the only one in which players needed 4 rounds to learn to cooperate[15] (see Appendix B). At the beginning, we observed chaotic behavior. Players did not cooperate in Round 1, they cooperated in Round 2, and again played non-cooperatively in Round 3 and 4 (see Appendix B). From Round 5 onwards, they cooperate regularly. Players' behavior in this auction is very similar to that of players who play sequences of finite Prisoner's dilemma supergames (Selten and Stoecker (1986)). Auction 11 was another exception. Already in Round 7, Player 1 defected; she did not make the promised side payments. In Round 8, she deviated again: Player 2, as selected winner, bid 58 but Player 1 bid 59 and transferred only few *soldi* to the other players (see Auction

---

[12]Cooperate as long as other players cooperate, otherwise do not cooperate.

[13]In fact, the total ex ante expected payoff is equal to 200 *soldi* when players always cooperate. Let $20 - x - 1$ be the number of rounds in which players cooperate. The expected payoff of a player who defects is roughly equal to 30 *soldi* in the round in which he deviates. After his defection, he gets, on average, 4.20 *soldi* in each of the remaining x rounds. That is, ex ante, he will defect for $x \leq 3.45$.

[14]We calculate these values considering only the sessions in which players reached an agreement, deviated from this agreement in some round and afterwards they stopped to cooperate. See Appendix B.

[15]Hereafter using "cooperate", we will mean an agreement which is always implemented.

11 in Appendix B). In Round 9, cooperation was restored. Player 2 and Player 3 were compensated for the loss they had suffered. Player 1 justified her behavior in Round 7 saying that she wanted to enjoy herself since the game was being played in a boring way (see Artale (1996)). Her explanation for her behavior in Round 8 was that she feared that Player 2 did not make any side payment to her for revenge. We will say something more about this "anticipated end effect" and more in general about the end effect in Section 2.7.

In what follows, we will refer to the 92% of all data in which all three players agreed on some collusive mechanism and implemented it.

In the remaining 5% of data, players bid non-cooperatively. They did it either because they did not reach any agreement (see above what we said about Auction 6), or because after a deviation by one player they did not want to cooperate anymore (auctions 5, 11, 16) or because they want to experience non-cooperative play in the last couple of rounds (Auction 15). In Subsection 2.4.3, we compare bids submitted by experimental subjects in this case with the unique symmetric equilibrium of a non-cooperative first-price auction.

There is only one case in which two players cooperated against another, in the last two rounds of the Auction 3. Players 2 and 3 cooperated against Player 1, who had to leave the room before they submit their bids[16]. In Subsection 2.4.4, we compare these data with the solution proposed by Marshall et al. (1994).

2.4.2   Observed Mechanisms

In this subsection, we describe mechanisms used by experimental subjects. The most frequently used mechanism was the *announcement mechanism*. According to the rule used to determine the sequence of announcements, we distinguish among three versions:

Simple announcement value mechanism: Players announced their values without caring about the order.

Random announcement value mechanism: Players used a random mechanism to choose in which order to announce.

Fixed order announcement value mechanism: Players specified the sequence of announcements so that Player i ($i = 1, 2, 3$) once announced as first one, once as second one and so on.

---

[16]This was the punishment inflicted to Player 1 by players 2 and 3 for having defected in Round 18. Player 1 did not make the promised side payments to Player 2 and Player 3.

Partition	Bids	Side Payments	
$51, \ldots, 57$	< 58	$0, \ldots, 0$	
$58, \ldots, 70$	58	$\frac{58-58}{3}, \ldots,$	$\frac{70-58}{3}$
$71, \ldots, 80$	59	$\frac{71-59}{3}, \ldots,$	$\frac{80-59}{3}$
$81, \ldots, 90$	60	$\frac{81-60}{3}, \ldots,$	$\frac{90-60}{3}$
$91, \ldots, 100$	61	$\frac{91-61}{3}, \ldots,$	$\frac{100-61}{3}$

Table 2.3: Lattice used by experimental subjects

The following splitting rule was common to the above mechanisms.
Splitting rule for announcement mechanisms: Subjects agreed that the player who announced the highest value $\hat{v}_{(1)}$ bid r in the legitimate auction and splitted equally the difference between his (announced) value and r, among all three players, i.e. he transferred $(\hat{v}_{(1)} - r)/3$ to each of both other players.

In Round 17 of Auction 1, players switched from the fixed order announcement mechanism to the first-price auction mechanism.
First-price auction: Players used a first-price auction to choose the winner. The player who made the highest bid $b_{(1)}$ bid r in the legitimate auction and splitted equally the difference between his bid (in the prior auction) and r, $(b_{(1)} - r)/3$.

The bid–bargain mechanism was used by experimental subjects in two auctions entirely and for ten rounds in Auction 9.
Bid–bargain mechanism: Players bargained making bids and asks. The player who made the highest bid bid r in the legitimate auction and paid each of the losers what he bid them or they asked him, respectively.

The lattice mechanism was used only in Auction 7.
Lattice mechanism: Players played in the main auction according to the lattice shown in Table 2.3 in order to choose the winner. They agreed that the winner of the legitimate auction had to make a side payment equal to his value minus his submitted bid divided by three.

Note that in all mechanisms, except in the bid-bargain mechanism, experimental subjects agree to divide equally the spoils among themselves[17].

---

[17]See Selten (1972) for a theory of "equal share".

Since players may bid and transfer only units of *soldi*, writing a/b (with a and b integer numbers) we mean always the integer part of a/b, $[a/b]$. In most of the sessions in which players used the announcement mechanism, the winner got the difference $a/b - [a/b]$. In Auction 7, on the contrary, players tried to distribute the difference $a/b - [a/b]$ equally through the rounds.

Note the difference between the first-price auction mechanism proposed by experimental subjects and the optimal first-price auction mechanism proposed by McAfee and McMillan (1992). According to the former, the winner pays each losing bidder the difference between his bid in the prior auction and the reserve price r divided by three. According to the McAfee and McMillan mechanism, on the contrary, the winner pays each of losing bidders the difference between his bid in the prior auction and the reserve price r divided by two.

Table 2.4 shows which mechanism was used in which auction. The most often used mechanism is the simple announcement mechanism. It was used in 11 auctions. Both the random announcement mechanism and the fixed order announcement mechanism have been used in one auction. The lattice mechanism was used once as well. The bid-bargain mechanism was used in two auctions entirely and in the last rounds of Auction 9. The first-price auction mechanism was used in the last three rounds of Auction 1. It would be worth noting that, in most of the cases, experimental subjects maintain the same mechanism when they cooperate. In Auction 1 , subjects switched once from one mechanism to another. In Auction 9, subjects switched twice from one mechanism to another; in Round 12, they switched from the simple announcement mechanism to the bid-bargain mechanism, and in Round 20 reversed back to the simple announcement mechanism[18].

We conclude this subsection by making two remarks. Firstly, note that the mechanisms we have described above are idealized. They are the fruit of discussions and of choices among different proposed mechanisms (see also Section 2.6). Secondly, note that, usually, players declared explicitly the rule of the mechanism. In few auctions, however, players did not explicitly and/or completely say which mechanism they were

---

[18]Actually, if we consider the announcement mechanism without side payments, i.e. a mechanism in which the player who announces the highest value obtains the object but *does not* make any side payment, as another mechanism, we should say that experimental subjects switched three times in Auction 9. They started with the announcement mechanism without side payments. In Round 2, they switched to the announcement mechanism with side payments and continued as we have seen above. In Section 2.6, we will say more about the announcement mechanism without side payments.

RA=random announcement mechanism
SA=simple announcement mechanism
FO=fixed order announcement mechanism
BB=bid-bargain mechanism
LA=lattice mechanism
FPA=first-price auction mechanism

A_1	A_2	A_3	A_4	A_5	A_6	A_7	A_8
RA/FPA	BB	SA	SA	SA	BB	LA	SA
A_9	A_{10}	A_{11}	A_{12}	A_{13}	A_{14}	A_{15}	A_{16}
SA/BB	SA	FO	SA	SA	SA	SA	SA

Table 2.4: Mechanisms distribution

going to use (Auctions 2, 6, 13, 15).

2.4.3 Winners' Bids and Payoff Shares

In this subsection we will show winners' bids and payoff shares. Without making any distinction among the different kinds of announcement mechanisms, Figures 2.1, 2.2 and 2.3 show the relative frequency of bids for announcement mechanisms, bid-bargain mechanism and lattice mechanism, respectively, when players cooperate. The fact that bids of 59 or 60 have a positive frequency in the announcement and the bid-bargain mechanism means that in some (but few) cases experimental subjects need a couple of rounds to realize that it is enough to bid the reserve price r to get the item. In all three rounds in which players used the first-price auction mechanism, they always bid r.

Figure 2.4 shows the payoff shares for the winner. The bottom line represents the payoff shares for the winner on average in case players equally split the difference between the highest *value* minus the reserve price. The top line represents the average payoff for the winner in case players use the mechanism proposed by McAfee and McMillan and bid according to Expression 2.1 (see Section 2.3 and Appendix B) in the prior auction. The line in the middle represents the experimentally observed average payoff shares for the winner. Note that when the bid-bargain mechanism is used, the payoff share for the winner is higher than in other cases. Note also that in some auctions in which players used the announcement mechanism, the payoff share for the winner is considerably higher than in the case in which subjects equally split the entire surplus. As we will see in Section 2.5, some players cheated announcing their values. This is the reason why the payoff share for the winner in some auctions is considerably higher than in the case in which subjects split all the surplus equally.

We conclude this subsection by comparing the highest bids submitted by experimental subjects when they bid non-cooperatively with the unique risk neutral Nash equilibrium (RNNE) (see Figure 2.5). Note that our results are different from the results obtained in the literature which report bidding substantially, and significantly, above the risk neutral Nash equilibrium (see Cox, Smith and Walker (1988); Dyer, Kagel and Levin (1989)). In particular, there is a tendency to underbid the RNNE in the first non-cooperative rounds, after having played a sequence of cooperative rounds. Observe, in particular, Auction 6. In the rounds preceding the cooperative phase, players bid above the RNNE; after 12 rounds of cooperation, they bid almost according to the RNNE (see Appendix D).

2.4.4   Two-Player Coalition versus One Individual Bidder

As we have already observed in Subsection 2.4.1, a two-player coalition played against a single player in the last two rounds of Auction 3. Table 2.5 compares the bids submitted by the coalition and by the single player with the theoretical prediction[19]. It is interesting to note that the coalition's bids are under the theoretical prediction and single player's bids over the theoretical prediction in both case. In particular, note that the single player risked to suffer a loss just to take revenge for having been excluded from cooperating in the last two rounds.

2.4.5   Summing Up

There are two important facts in this subsection that are worthy to be emphasized. Firstly, experimental subjects agreed, in most of the cases, on a very simple rule to implement cooperation and played accordingly. Secondly, in most of the cases, they declared themselves for an equal division of the difference between the highest *revealed* value and the reserve price r.

---

[19]Details about the solution are given in Section 5.2.

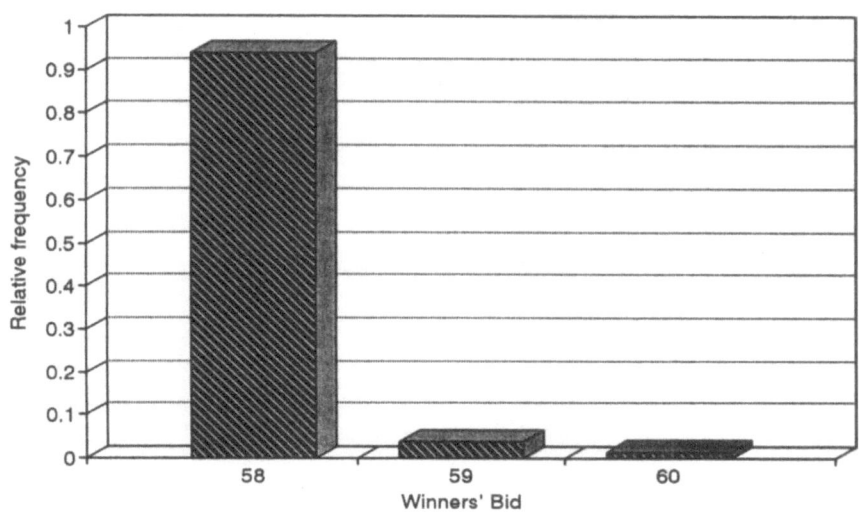

Figure 2.1: Winners' bid in the Announcement mechanism

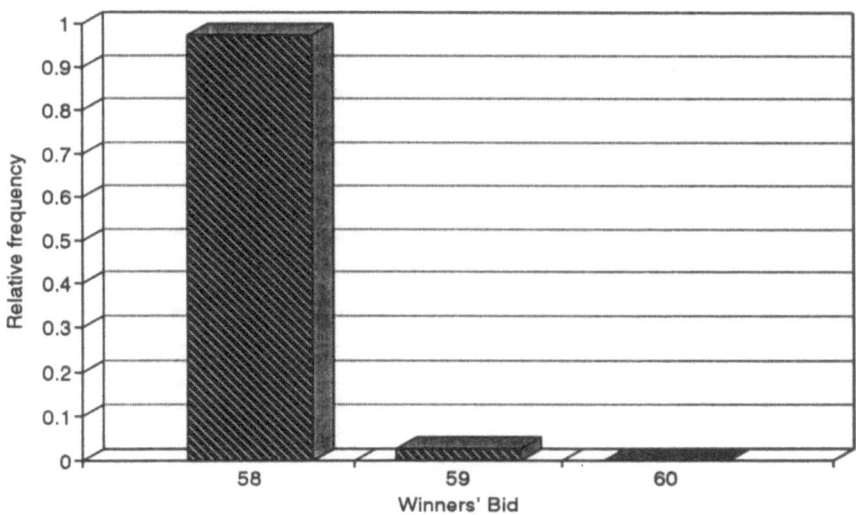

Figure 2.2: Winners' bid in the Bid-bargain mechanism

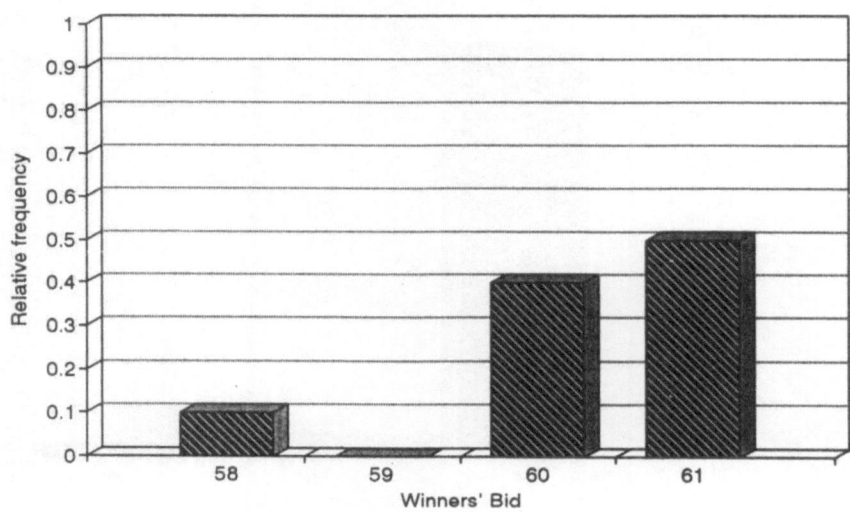

Figure 2.3: Winners' bid in the Lattice mechanism

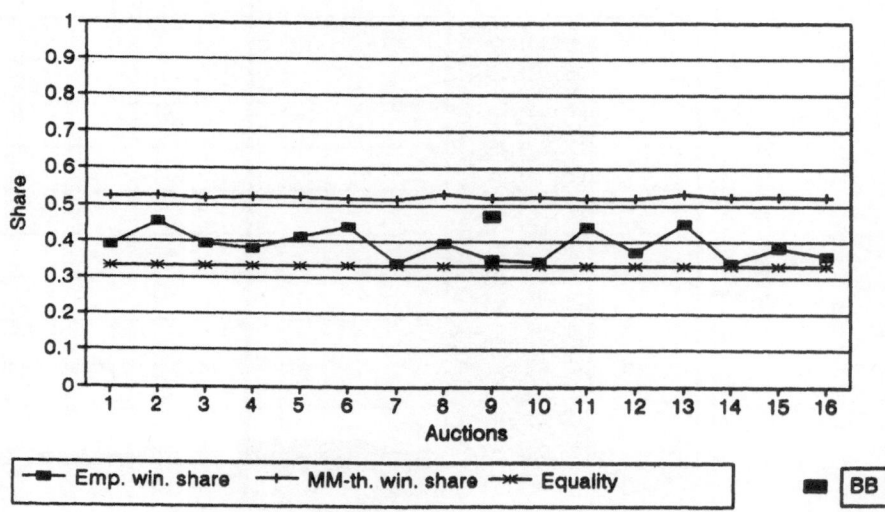

Figure 2.4: Winners' payoff share

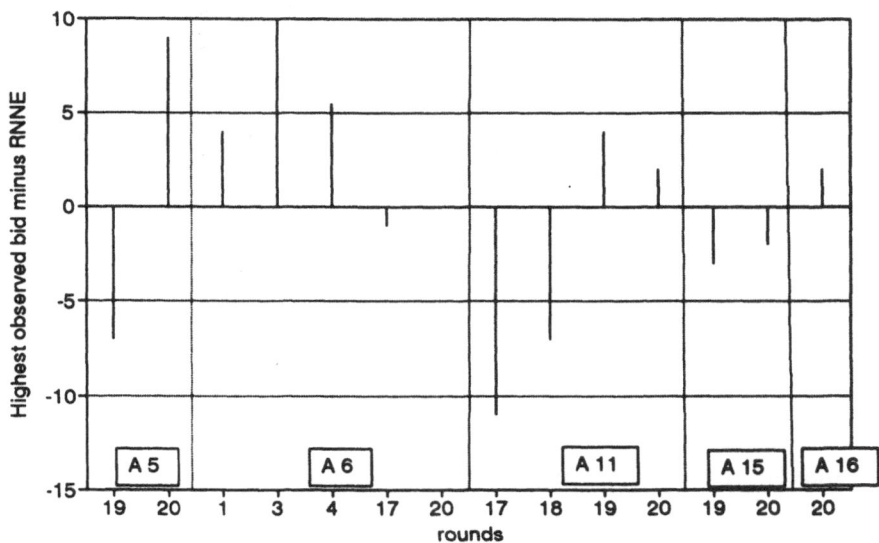

Figure 2.5: Highest observed bid minus RNNE

$v_C = \max\{v_2, v_3\}$: coalition's value
$v_S = v_1$: single player's value
b_C: coalition's bid
b_S: single player's bid
b_C^*: coalition's equilibrium bid
b_S^*: single player's equilibrium bid

Round	v_C	v_S	b_C	b_S	b_C^*	b_S^*
19	72	51	64	60	66.5	51
20	83	76	70	70	72.3	67.5

Table 2.5: Coalition versus Single Player (Auction 3)

2.5   Theoretical Predictions versus Data

In this section, we compare the experimental observations of each mechanism with the theoretical prediction of that mechanism. A rigorous strategic analysis of each mechanism is developed in the Chapter 4. Here, we rely on the reader's intuition.

We ask the following question. Given that players agreed on some mechanism and played accordingly, which equilibrium has this mechanism?

We will distinguish among four different mechanisms. In Subsection 2.5.1, we start to analyze the announcement mechanism, in which players made their announcements sequentially. As we will see, experimental subjects did not play according to the theoretical prediction. In most of the cases, they announced their true values. In Subsection 2.5.2, we focus on and try to explain the behavior of players who cheat. Subsection 2.5.3 reports the expected payoff that players obtain when they all play according to the theoretical prediction, when two announce their true values and another announces his best reply, and when all three announce their true values. We compare these payoffs with the observed average payoff, and with the payoff of an optimal mechanism. Subsection 2.5.4 analyzes the bid-bargain mechanism, i.e., the mechanism in which players made bids and asks. Modeling players' behavior, we consider only the bids. As we will show in Subsection 2.5.5, the bid-bargain mechanism is almost optimal, i.e. the probability that the player with the highest value does not get the item is very low. We compare the expected payoff which players obtain when they play according to the model in equilibrium with the observed average payoff. Subsection 2.5.6 investigates the lattice mechanism, in which players bid in the legitimate auction using the lattice shown in Table 2.3. Given the side payments proposed by experimental subjects, we find the partition of equilibrium. Afterwards, given the partition proposed by experimental subjects, we find the side payment function of equilibrium. Expected and the observed average payoffs for the lattice mechanism are compared in Subsection 2.5.7. The first-price auction mechanism, in which players play a prior first-price auction before they play the legitimate auction, is analyzed in Subsection 2.5.8. Finally, Subsection 2.5.9 sums up the most important facts.

2.5.1 The Announcement Mechanism

As we have seen above, the announcement mechanism was the most often used mechanism. Players who use this mechanism announce their (possibly true) values in sequence. The player who makes the highest announcement gets the item, pays the reserve price to the auctioneer and pays each losing bidder the difference of his announcement and the reserve price, divided by three. It was used in three different versions. Each version determines the sequence of announcements. The simplest one leaves players free to make their announcement in the sequence they want. The other two versions make use of a random device and of a fixed order, respectively. Now, apart from different kinds of announcement mechanism, that is apart from the fact whether any sequence of announcements is determined, which is the best announcement for the player who announces as the first one? And which is the best announcement for the player who announces second or third, respectively? Here we give only an intuition of the solution which is provided in detail in Chapter 4.

We begin with the player who announces last. He has heard the announcements of the other two players. If his value is higher than the highest announcement he has heard, then he should announce the highest announcement plus one[20]. If he has a value which is equal to, or lower than, the highest one he has heard, then any announcement equal to or higher than the highest announcement he has heard is a dominated strategy[21]. Since he is indifferent between announcing his own value and announcing any value lower than the highest announcement he has heard, we can say that a best reply for the player who announces as the third one is announcing the highest announcement plus one, when his value is higher than the highest announcement he has heard, and announcing his own value, when this is equal to, or lower than, the highest announcement he has heard.

Now, we consider the player who announces as the second one. He has heard the announcement of the preceding player and he will anticipate the behavior of the player who announces last. If the first announcement is lower than his own value he will overbid if he has a relatively low value

---

[20]This is not always true. Consider this example. The value of the player who announces last is equal to 58 and the highest announcement he has heard is equal to 52. In this case, he would be indifferent among announcing any value in {58, 59, 60}. See Section 4.2 for more details.

[21]Actually, this is not always true. Consider this example. Let the value of the player who announces last be equal to 80 and let the highest announcement he has heard be also equal to 80. In this case, his best announcement is equal to 81. Other details are provided in Section 4.2.

and he will underbid if he has a relatively high value. The intuition be-
hind this solution is easy to understand. Since the player who announces
as the second one anticipates the behavior of the player who announces
last, when he has a relatively low value he will announce more than his
value, hoping that the player who announces last has value higher than
the value he has announced. Let us now consider the case in which a
player has a relatively high value. In this case, if he announces his own
value, the probability to get the item is high. If his announcement is
slightly lower than his own value then the probability to get the item is
slightly reduced but the potential gain increases. Since the players are
assumed to be risk neutral, they maximize the sum of the product of
probabilities to get the item times the gains. If the first announcement
is equal to or higher than his own value, he will be indifferent between
announcing his own value and any other value which is lower. Therefore,
we can say that he will announce his own value.

Finally, we consider the player who announces as the first one. He will
anticipate the behavior of the succeeding player. When he has relatively
low values he will overbid, since the probability that the other two players
have values higher than his announcement is high. If, however, he has
relatively high values he will underbid, since the probability that the
other two players have values higher than his announcement is low[22].

Figure 2.6 plots the theoretical prediction for the player who announces as the first one (see Proposition 1 in Section 4.2) and experimental
observations on average. Two facts are immediately observable. Firstly,
players who have low values did not overbid. They did not anticipate the
behavior of players who announce afterwards. Secondly, players started
to cheat from a value of 81 (and in a more resolute way, from 91) but,
on average, they overbid the theoretical prediction.

Figures 2.7 and 2.8 plot theoretical predictions (see Proposition 1
in Section 4.2) versus average experimental observations for the player
who announces second and third, respectively, conditional on what they
heard. The forty-five degrees line represents the theoretical prediction,
conditional on what the player heard. That is, it represents the values
that the player who announces second and third, respectively, is expected
to announce given his own value and the announcement(s) he has heard.
For each best announcement, we consider the average of the observed
announcements. Note that these figures do not make any distinction
among theoretical predictions which are conditioned on different announ-

---

[22]We have disregarded the restriction given by the units. Remember, in fact, that
experimental subjects may transfer only units. However, the theoretical solution given
in Figures 2.6 until 2.8 considers this restriction.

cements. Note also that we do not have experimental observations for each theoretical prediction.

Neither the player who announces second nor the player who announces third played approximately according to the theoretical prediction[23]. For these two positions, there is a clear tendency of experimental subjects to overbid the theoretical prediction. We test the hypothesis H_0 :"All three players play according to the theoretical prediction more often than not" with a one-tailed binomial test for each auction separately[24]. Since new values are drawn in each round, we consider each round as an independent observation. If all three players play according to the theoretical prediction we attribute the value one to the round, otherwise we attribute the value zero. Doing so, we can reject our hypothesis with a one-tailed binomial test at significance level $p = .001$.

Remark One might object that this model does not fit subjects behavior well in case they use a simple announcement mechanism. The reason is that players' (ex ante and interim) expected payoffs depend on which position they announce. That is, the sequence of announcements is a strategic component of this model. Actually, this criticism does not apply to the way in which we model experimental subjects' behavior for the following cogent reasons.

Firstly, there is no statistical evidence of a correlation between values and position. Some groups of players who use the simple announcement mechanism are aware of the fact that players with high values may want to hear the other two players before they announce, since they try to change the sequence of announcement in each round (see Auctions 4, 8 and 16 in Appendix C). In order to see whether there is a relation between the position in which players announce and the rank of values, we calculated a gamma statistic G (see Siegel and Castellan (1988)) for each auction in which players used the simple announcement mechanism[25]. In

---

[23]Data used to produce Figures 2.6, 2.7 and 2.8 are reported in Appendix C.

[24]We have to test each auction separately, since results depend on group composition.

[25]Consider the variables P and v. The first variable, P, represents the position in which a player may announce. It can take on the values $1, 2, 3$. The second one, v, represents the value that a player can draw, and it takes on the values $51, \ldots, 100$. The parameter γ is the difference in the probability that within a pair of observations P and v are in the same order and the probability that within a pair of observations P and v disagree in their ordering, provided that there are no ties in the data. That is,

$$\gamma = \frac{Prob\{P\&v \text{ agree in order}\} - Prob\{P\&v \text{ disagree in order}\}}{1 - Prob\{P\&v \text{ are tied}\}}$$

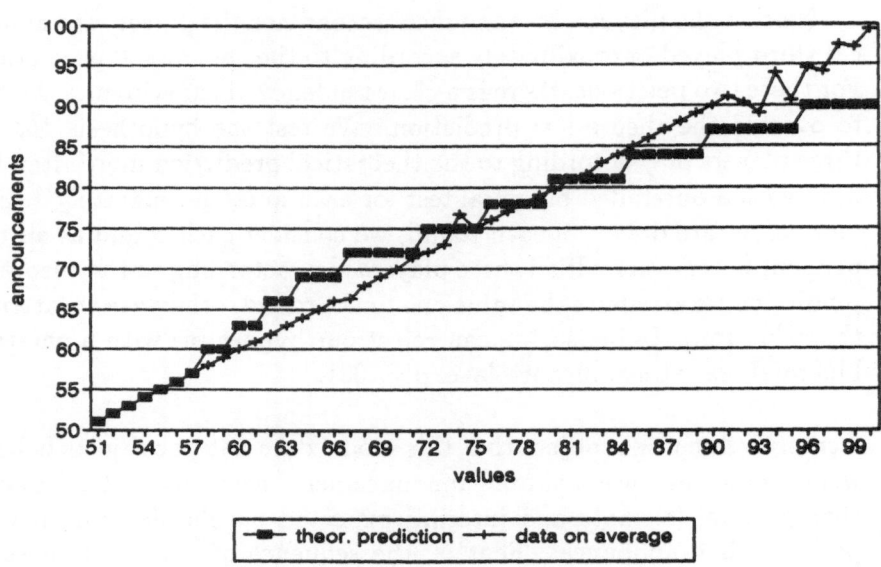

Figure 2.6: Theoretical prediction versus experimental observations on average (Position 1)

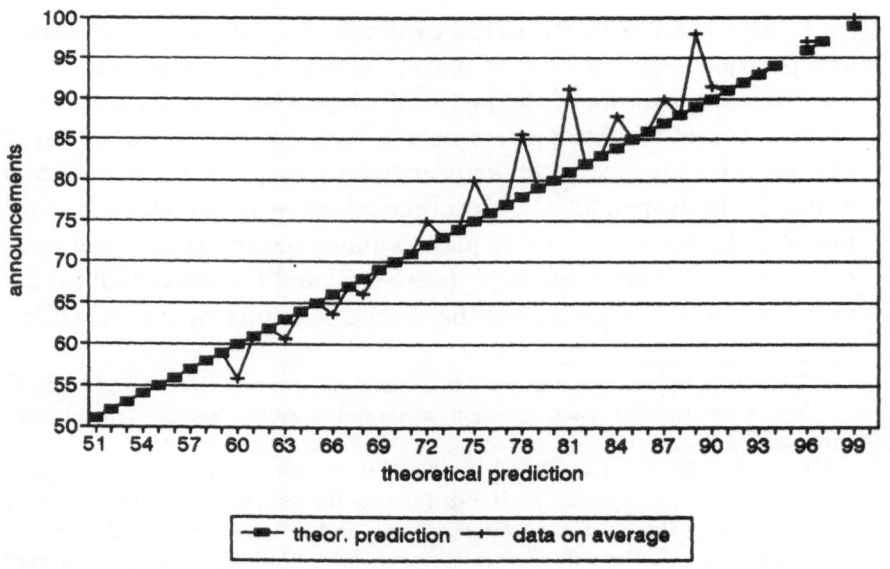

Figure 2.7: Theoretical prediction versus experimental observations on average (Position 2)

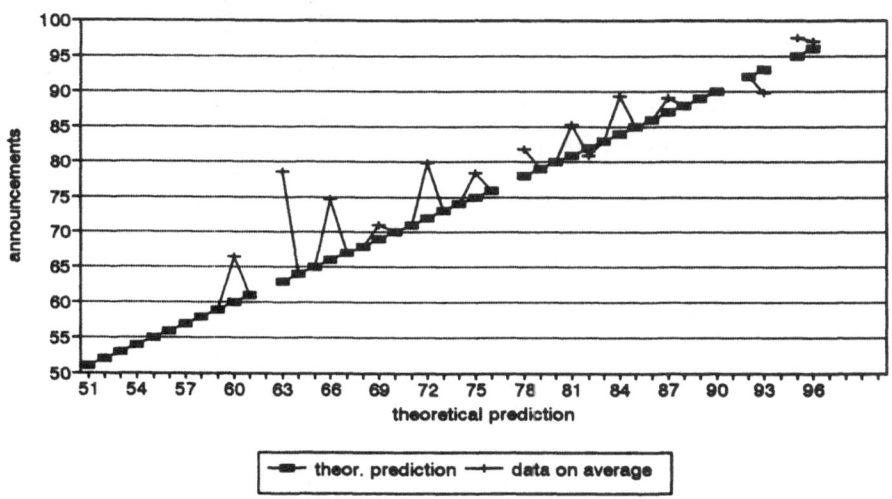

Figure 2.8: Theoretical prediction versus experimental observations on average (Position 3)

no case did we get a significant result testing the hypothesis $H_0 : \gamma = 0$ against the hypothesis $H_1 : \gamma \neq 0$ for $\alpha = 0.01$. In Table 2.6, we report the values of the gamma statistic G and the value of z in brackets[26].

Secondly, even if we had observed such a correlation we could not have used any other non-trivial model. Using this model, we do not

Since we do not know the probabilities in the population, we must estimate them from the data. Thus, we use the statistic G to estimate γ. See Castellan and Siegel (1988) for a definition of G.

[26]For $\alpha = .01$, we have $z = 2.58$ for a two tailed test.

G_3	G_4	G_5	G_8	G_9	
-0.26	0.168	-0.070	0.046	0.23	
(-0.94)	(0.63)	(-0.25)	(0.18)	(0.712)	
G_{10}	G_{12}	G_{13}	G_{14}	G_{15}	G_{16}
-0.1	0.07	0.062	-0.51	-0.25	-0.071
(-0.37)	(0.26)	(0.224)	(-2.25)	(-0.911)	(-0.255)

Table 2.6: Values of the gamma statistic G

want to *describe* experimental subjects' behavior but on the contrary we
want to *prescribe* how rational players should play and use this model as
benchmark for our analysis.

2.5.2   Announcement Mechanism: The Behavior of Players Who Cheat

As we have seen in the last subsection, experimental subjects do not play
on average according to the theoretical prediction. Now, we focus only
on players who cheat. We want to find out whether players who cheat
play (more or less) according to the theoretical prediction.

The number of players who cheat at least once is equal to 14. Among
these 14 players, not everybody cheats systematically. That is, there are
players who cheat when they announce in position two or in position three
but not in position one (having the same value)[27]. Similarly, there are
players who cheat only when they have high values irrespective of their
position. Figures 2.9, 2.10 and 2.11 show the relative frequency of the
difference of value minus announcement for players who have cheated at
least once. Zero means that a player announced his own value, although
he cheated at least once and he had the possibility to cheat[28]. Note
that players never overbid their values. It is interesting to observe that
the frequency of zero for players who announce in position two as well
as for players who announce in position three is surprisingly high (0.18
and 0.28, respectively).   This fact could induce us to think that the
players' announcement, when they cheat, does not depend on their own
value and, eventually, on other players' announcement, but perhaps on
how frequently they cheated in the past rounds. One could think that
some players who cheat have some dynamic strategy in mind. They cheat
occasionally because they fear that other players can realize that they
are cheating and be offended or disappointed. In any case, they fear that
cooperation could break down.

To test this conjecture we define an index, RI, the rate of inconsistency. Roughly speaking, we say that if a player cheated already in
position 1, he is expected to cheat in position 2 or 3 as well if he has
a value that is at least equal to or higher than the value he had, when
he cheated in position 1 or 2. More formally, we say that a player is
inconsistent if, after having cheated in position l, $(l = 1, 2)$ he does not't
do it in position $i > l$ (given that he can), having a value $v_i \geq v_l$.

---

[27]See Appendix C.

[28]Obviously, a player who announces in position one has always the possibility to
cheat, since he has not yet heard any previous.

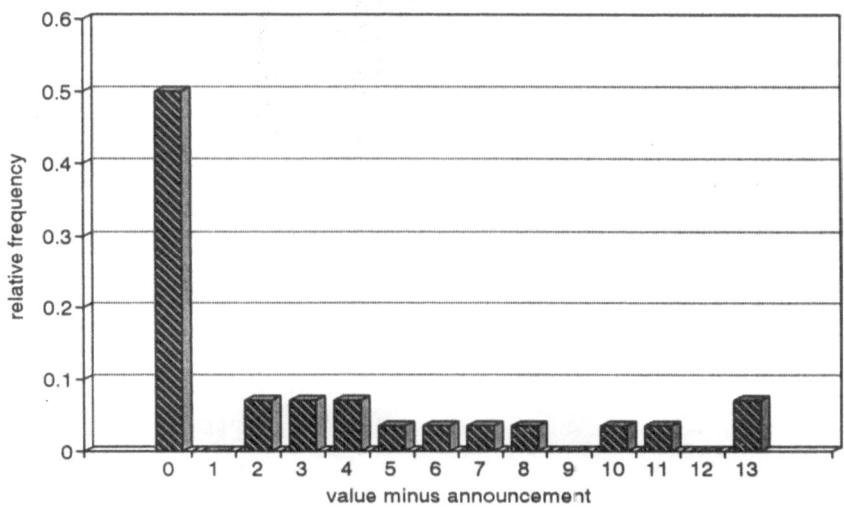

Figure 2.9: Relative frequency of $v_1 - \hat{v}_1$

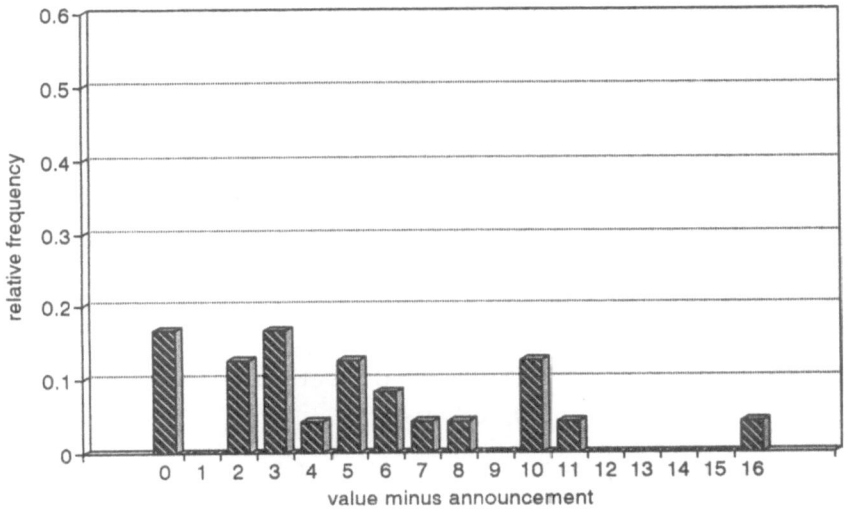

Figure 2.10: Relative frequency of $v_2 - \hat{v}_2$

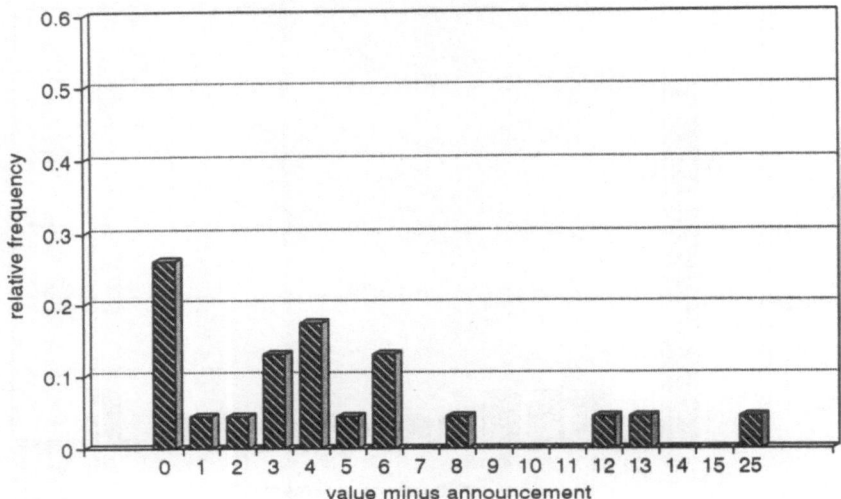

Figure 2.11: Relative frequency of $v_3 - \hat{v}_3$

Let $\underline{v_j} = \min\{v_{j_1}, \ldots, v_{j_t}\}$ be the lowest value that a player has, cheating in position j in Round s, $1 \leq s \leq t$. Let us define the following two events:

$E_t = \{$ subject k $(k = 1, 2, 3)$ cheats in round t for the first time in position i, or he has already cheated in position l having a value $\underline{v_l}$, and he is now in position $i \geq l$ and has a value $v_i \geq \underline{v_l}$ $\}$

$O_t = \{$ subject k $(k = 1, 2, 3)$ cheats in round t $\}$

Let I_{E_t} and I_{O_t} be indicator functions of the events E_t and O_t respectively, that is

$$I_{E_t} = \left\{ \begin{array}{ll} 1 & \text{if } E_t \text{ is true} \\ 0 & \text{otherwise} \end{array} \right.$$

$$I_{O_t} = \left\{ \begin{array}{ll} 1 & \text{if } E_t \text{ and } O_t \text{ are true} \\ 0 & \text{otherwise} \end{array} \right.$$

We can now define the rate of inconsistency RI

$$RI = 1 - \frac{\sum_t I_{O_t}}{\sum_t I_{E_t}}$$

Note that $0 \leq RI < 1$. We will say that a player is not inconsistent when RI is close to zero. If, otherwise, RI is close to one, we will say that he is inconsistent[29]. Table 2.7 shows the rates of inconsistency in detail for each player. From the value of $\overline{RI} = (\sum_{j=1}^{14} RI_j)/14 \simeq 0.06$ we cannot conclude that players are inconsistent. This result can at least explain Figures 2.9 to 2.11. A player who cheats in position 3 may not want to cheat in position 1 or 2, or he may not want to do it if he has not drawn a very high value. Similarly, a player may want to cheat in position three only when he has an high incentive.

Now, we restrict ourselves to the data of players who cheat when they actually do it. We want to find out how they cheat. Figures 2.12, 2.13 and 2.14 compare average theoretical predictions with average experimental observation, for players who announce in position one, two and three, respectively. Even in this case, we cannot conclude that experimental subjects cheat according to the theoretical prediction. This discrepancy can be explained by the following arguments. Firstly, experimental subjects cheat only when they have high values. On average, a player who cheats in one of three positions must have at least a value of 81. This consideration is important to understand the behavior of players who announce in position one and in position two. Secondly, experimental subjects need a high monetary incentive to cheat. Consider the following example. Let us assume that the player who announces in position three has drawn a high value, say 97, and that the highest announcement he has heard was 93. Now, if he were playing according to the theoretical prediction, he should announce 94. But the monetary incentive to announce 94 is very low (+3). Moreover, announcing 94 he could make the other two players suspicious that the third one is cheating. Finally, another explanation of the discrepancy between data and theoretical predictions could be that experimental subjects do not per-

---

[29]Note that this definition cannot distinguish between players who cheat say two or three times at the beginning in position one or two and stop cheating thereafter (which we call unintentionally inconsistent), and players who, having cheated say once in position one do not cheat in position three but cheat again in position two (which we call intentionally inconsistent).

RI_1	RI_2	RI_3	RI_4	RI_5	RI_6	RI_7	RI_8
0	0	0,16	0	0,3	0,1	0	0

	RI_9	RI_{10}	RI_{11}	RI_{12}	RI_{13}	RI_{14}
	0	0	0	0	0	0

Table 2.7: Rates of inconsistency

ceive each round as a one-shot game (as the theoretical prediction does) but as a repeated game[30]. When they cheat there is a positive probability to be discovered, and this could compromise the cooperation in the next rounds.

We explain the amount of cheating of experimental subjects just looking at the value of each player who cheats and the highest value he has heard. We propose the following linear equations to describe the announcements of Player 1[31], 2 and 3, when they cheat[32]

$$\hat{v}_1 = \beta_1 \cdot v_1$$

$$\hat{v}_2 = \beta_2 \cdot v_2$$

$$\hat{v}_3 = \beta_{31} \cdot v_3 + \beta_{32} \cdot \hat{v}_m$$

where $\hat{v}_j$ and v_j are, respectively, the announcement and the value of the player who announces in position j, and $\hat{v}_m = \max\{\hat{v}_1, \hat{v}_2\}$. We have estimated the value of the betas using the standard OLS method.

Table 2.8 reports results obtained. Consider the third relation. It says that the announcement of Player 3 is almost a convex combination of his own value and the highest announcement he has heard with weights 0.5309 and 0.473, respectively. Relying on this result, we have repeated the regression of the third relation subject to the constraint that β_{31} and β_{32} are weights of a convex combination (see Table 2.9). The results tell us that Player 3's announcement (when he cheats) is a convex combination of his own value and the highest announcement he has heard. Remember that the theoretical prediction says, that in this case, $\hat{v}_m + 1$ should be announced. That is, the theoretical prediction would say that $\beta_{32} \to 1$ and $\beta_{31} \to 0$.

This model describes how experimental subjects cheat when they actually do it. To explain their cheating behavior more carefully, we also need to know when they cheat. In Chapter 3, we will consider a more general descriptive model and we will estimate the probability that a player cheats.

2.5.3   Announcement Mechanism: Optimality

The model proposed as benchmark of analysis has an equilibrium which does not always ensure the optimal allocation, that is the allocation in

---

[30]See Rubinstein (1991) for such a distinction.

[31]When it is not confusing, we will write Player i instead of player who announces in position i.

[32]Note that the verb *describe* is used without pointing out that these equations constitute a model which explains players' behavior.

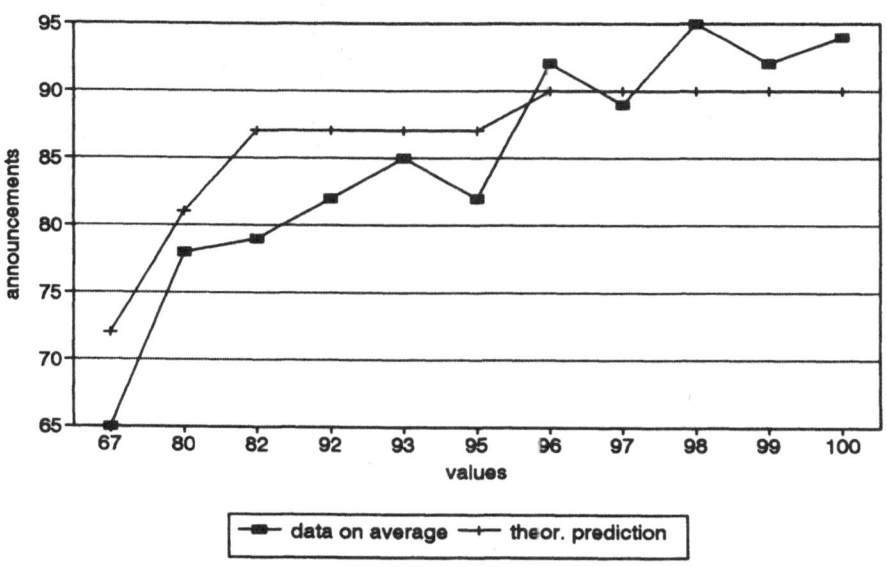

Figure 2.12: Theoretical prediction versus exp. observations for a player who cheats in Position 1

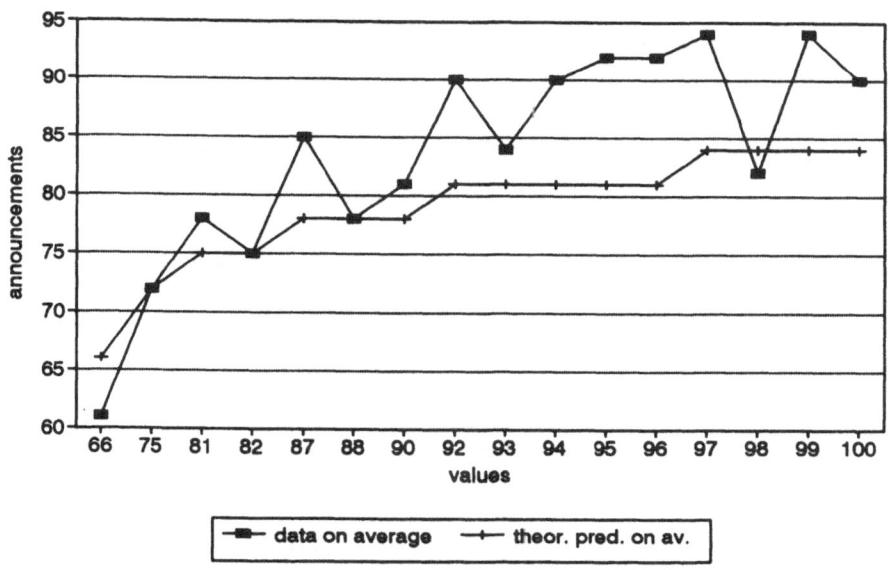

Figure 2.13: Theoretical prediction versus exp. observations for a player who cheats in Position 2

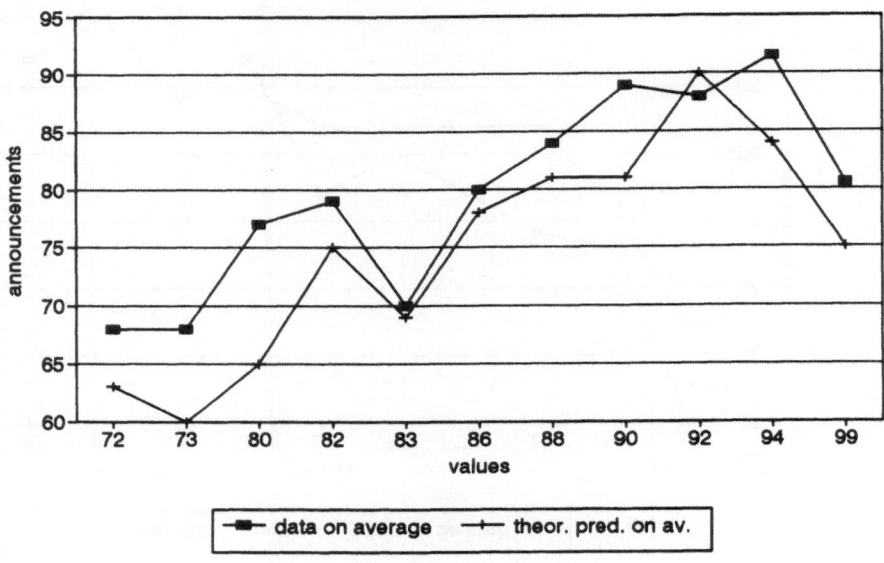

Figure 2.14: Theoretical prediction versus exp. observations for a player who cheats in Position 3

which the player with the highest value gets the item. Note however that in 95% of the cases, players who play according to the announcement mechanism allocate the item optimally (see Table 2.10). This is due to the fact that players play *naïvely* in most of the rounds. That is, players announce their own value instead of cheating.

We have simulated the ex ante payoffs for the case in which two players play naïvely and one player plays his best reply[33], and for the case in which all three players play a particular equilibrium of this mechanism

---

[33]The position of the player who plays the best reply is randomly drawn in each round.

	value of $\hat{\beta}$	t	Sig. level	R^2
β_1	0.9278	87.18	0.0001	0.998
β_2	0.933	103.46	0.0001	0.9982
β_{31}	0.5309	5.21	0.001	0.997
β_{32}	0.473	3.93	0.0012	0.997

Table 2.8: OLS regression results

$\hat{\beta}_{31}$	$\hat{\beta}_{32}$	F	*Sig. level*	R^2
0.51234	0.48766	0.031699	0.0000	0.9768

Table 2.9: Constrained OLS regression results

(see Proposition 1). Of course, when all three players play naïvely, the ex ante expected payoff is equal to the optimal payoff. We report these values below in Table 2.11. Note that the observed average payoff is practically equivalent to the optimal payoff.

2.5.4 Bid-bargain Mechanism

This mechanism was used in two auctions entirely and for nine rounds of Auction 9. Players who used this mechanism bargained making bids and asks. The bargaining process started usually with a bid followed by other bids and, sometimes, by asks. Since all players who made asks had low values (60 on average) in most of the cases, one can reasonably assume that they did it trying to get more from other players. In the model that we use as a benchmark, we do not explicitly model asks. We assume that players submit their bids simultaneously to make our problem more tractable. Moreover, the solution of this problem, (in which bidders bid simultaneously), should not be very different from the solution of the problem in which players bid sequentially.

In what follows, we illustrate the model used as benchmark and we give an intuition of the solution. Formal results are derived in Chapter 4. Let us imagine that players use a thermometer. The thermometer has only 14 numbers (from 1 to 14) and at the beginning it displays 1. At b, ($1 \leq b \leq 14$), all three players have to announce simultaneously "Yes" or "No". If at least two players say "Yes" the temperature goes up to $b + 1$, otherwise if all three players say "No", the winner (who is chosen randomly) has to pay 58 to the auctioneer and $b - 1$ to each loser, and

A_1	A_3	A_4	A_5	A_8	A_9	A_{10}	A_{11}	A_{12}	A_{13}	A_{14}	A_{15}	A_{16}
20	20	20	17	20	19	20	17	18	20	20	17	19

Table 2.10: Number of rounds in which the player with the highest value got the item playing the announcement mechanism

the round ends. If only one player says "Yes", he gets the item, pays b to each loser, and the round ends. Note that bids higher than 14 are not serious since the highest value is 100. Saying "No" at 15, a player who has drawn 100 gets 14 for sure. On the contrary, saying "Yes", he would get only 12 (that is, 100-58-30).

Now, given that a player has a value v, which is the highest bid he will submit in equilibrium? Let $b^h(v)$ be the highest bid until which a player with value v is willing to say "Yes", given that at least another player is active. The equilibria we are looking for are symmetric equilibria in pure strategies. That is, given that $(n-1)$ players bid according to some $b^h(v)$, it is also optimal for the nth player to play according to the same $b^h(v)$. In Chapter 4, we compute one of these equilibria through iterations of the best reply. We will show that our model seems to have two equilibria. One of these equilibria seems intuitive to be selected, namely, the equilibrium which is calculated starting from $[\frac{v-58}{3}]$, the integer part of $\frac{v-58}{3}$, for each v.

In Figure 2.15 and 2.16, we have plotted the highest equilibrium bid versus the highest observed bid on average and the highest bid in the case that players bid *naïvely*, i.e., until $b = [(v-r)/3]$, versus the observed highest bid on average, respectively. The forty-five degrees line represents the theoretical prediction (see Section 4.3). This is calculated assuming that all three players play according to the theoretical prediction. For each equilibrium highest bid (for each naïve highest bid, respectively), we consider the highest bids explicitly made by experimental subjects. We do not consider the asks. Unfortunately, the meager number of data does not allow us to clearly identify features of players' behavior. Nevertheless, it seems to us that there is a tendency to overbid the theoretical prediction for values of $b^h \leq 6$ and to underbid it for values of $b^h \geq 7$. There are two exceptions. In Figure 2.15, players underbid at $b^h = 4$ and overbid at $b^h = 9$. In Figure 2.16, players play according to the theoretical prediction at $b^h = 2$ and overbid at $b^h = 9$. Note that both the equilibrium highest bid and the naïve highest bid fit the data badly at low values of b^h. This is due to the fact that, at low

Optimal	Naïve	Best Reply	Equilibrium	Observed Payoff
10.04	10.04	11.67	8.84	10.09

Table 2.11: Ex ante expected and observed payoffs for the announcement mechanism

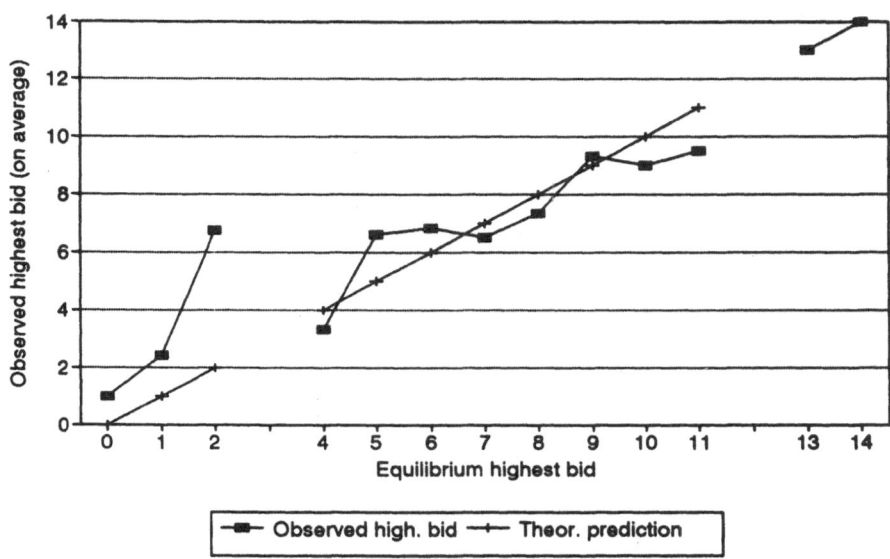

Figure 2.15: Equilibrium highest bid versus observations on average

bids, players do not increase their bids in single units but they jump. Moreover, sometimes, they make high bids (that they could not pay) in order to push up the highest bid (see Artale (1996)). Figures 2.17 and 2.18 plot the relative frequency of the deviation of the data from the equilibrium highest bid and from the naïve highest bid, respectively. We consider the distribution of the differences between the observed highest bids and their associated equilibrium highest bids and the distribution of the differences between the observed highest bids and their associated naïve highest bids, respectively. We have calculated the mean and the variance of these two distributions and we have found that they are 0.28 and 5.6 for Figure 2.17, and 0.92 and 5.28 for Figure 2.18, respectively. Note that, on average, experimental subjects overbid the theoretical prediction in both cases, but the equilibrium highest bid seems to fit the data better than the naïve highest bid.

2.5.5   Bid-Bargain Mechanism: Optimality

The bid-bargain mechanism model is almost optimal, it always selects as winner the player with the highest $b^h(v)$. We say "almost" optimal instead of optimal because $b^h(v)$ is not strictly monotonically increasing (see Figures 4.2 and 4.4). In case of ties (which have a positive probability in our model), there is a positive probability that the bidder with

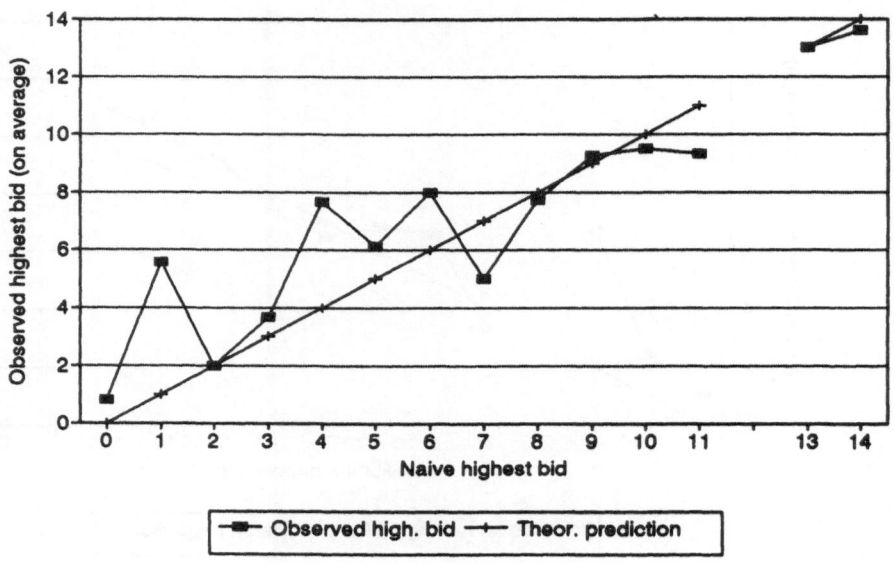

Figure 2.16: Naïve highest bid versus observations on average

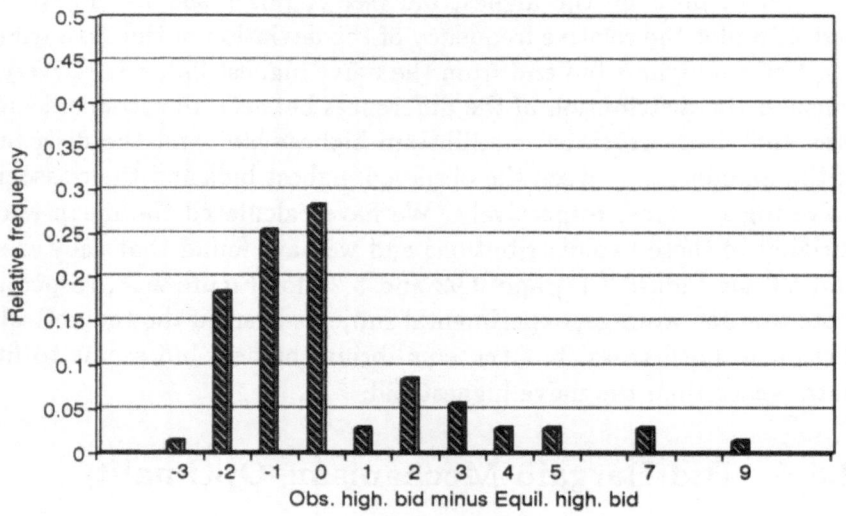

Figure 2.17: Observed highest bid minus equilibrium highest bid

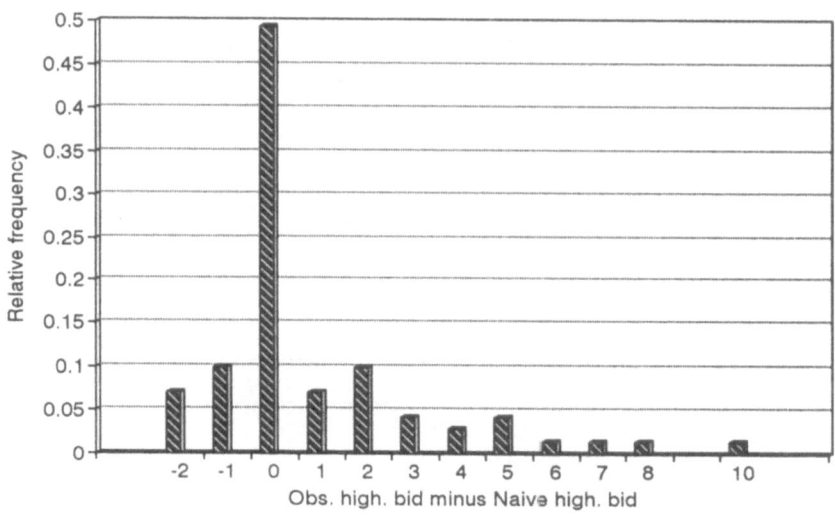

Figure 2.18: Observed highest bid minus naïve highest bid

the highest value does not obtain the item[34]. Note that in this model non-optimality is due to integer numbers. In a continuous version of this model, we have optimality since the probability of ties is zero. In the announcement mechanism, on the contrary, non-optimality persists even if we adopt a continuous version.

In our experiment, the item was always allocated optimally when players cooperated (see Appendix B). Table 2.12 shows the ex ante payoffs when all players play naïvely, when two of them play naïvely and the other one plays the best reply, when all three players play in equilibrium and the observed average payoff.

---

[34]However, the probability to get a non-optimal result is low, it is just 0.038.

Optimal	Naïve	Best Reply	Equilibrium	Observed Payoff
10.04	9.65	10.22	10.002	11.4

Table 2.12: Ex ante expected and observed payoffs for the bid-bargain mechanism

Partition	Bids	Side Payments
$51, \ldots, 57$	< 58	0
$58, \ldots, 73$	58	0
$74, \ldots, 86$	59	$\frac{71-59}{3}$
$87, \ldots, 100$	60	$\frac{81-60}{3}$

Table 2.13: Best Reply Lattice

Partition	Bids	Side Payments
$51, \ldots, 57$	< 58	0
$58, \ldots, 72$	58	0
$73, \ldots, 86$	59	$\frac{71-59}{3}$
$87, \ldots, 100$	60	$\frac{81-60}{3}$

Table 2.14: Equilibrium Lattice (given the side payment function proposed by experimental subjects)

2.5.6   Lattice Mechanism

This mechanism was used only in one auction. Players who chose this mechanism fixed and played according to the lattice shown in Table 2.3. Players submitted always their bids according to this lattice and made their side payments according to the lattice too. They never cheated (see Auction 7 in Appendix C). Now, given that two players play according to the lattice, is it a best reply for another player playing according to the lattice as well? The answer to this this question is no, as Table 2.13 shows. Note that calculating the best reply, we assume that a player, bidding $b, (b \in \{58, \ldots, 61\})$ and getting the item, will pay each of losing bidders always the minimum of the respective interval. Iterating the best reply we can find the equilibrium shown in Table 2.14. Now, we can ask the following question. Given the partition proposed by experimental subjects, does there exist a side payment function such that this partition with these side payments is an equilibrium? The answer to this question is affirmative as Table 2.15 shows. In Chapter 4, we describe how one can find such side payment function.

Partition	Bids	Side Payments
$51, \ldots, 57$	< 58	0
$58, \ldots, 71$	58	0
$72, \ldots, 81$	59	$\frac{70-59}{3}$
$82, \ldots, 90$	60	$\frac{77-60}{3}$
$91, \ldots, 100$	61	$\frac{84-61}{3}$

Table 2.15: Equilibrium Lattice (given the partition proposed by experimental subjects)

Optimal	Naïve	Best Reply	Equilibrium	Observed
10.04	9.12	9.52	9.16	9.73

Table 2.16: Ex ante expected payoffs for the lattice mechanism

2.5.7 Lattice Mechanism: Optimality

The lattice mechanism is a sub-optimal mechanism. Using the partition adopted by experimental subjects, we calculate the probability that the player with the highest value does not obtain the item. When players play according to the equilibrium shown in Table 2.14, this probability is 0.14. One can reduce this probability increasing the number of elements of the partition. In this way, each subset of the new partition has a smaller number of values, and the probability to draw all three values from the same subset is lower than in the old partition. Doing so, however, the expected ex ante payoff of each player decreases. In Chapter 5, we will show that when continuous bids are allowed one can obtain optimality, that is, the player with the highest value gets the item. However, we will see that a small waste of money persists.

In our experiment, a dice was thrown in five rounds to choose the winner, and in two of these five rounds, the bidder with the second highest value got the item. Table 2.16 shows the simulated ex ante expected payoffs when players play naïvely (i.e. according to the proposed lattice), when two players play naïvely and another one plays the best reply, and when all three players play according to the equilibrium shown in Table 2.14.

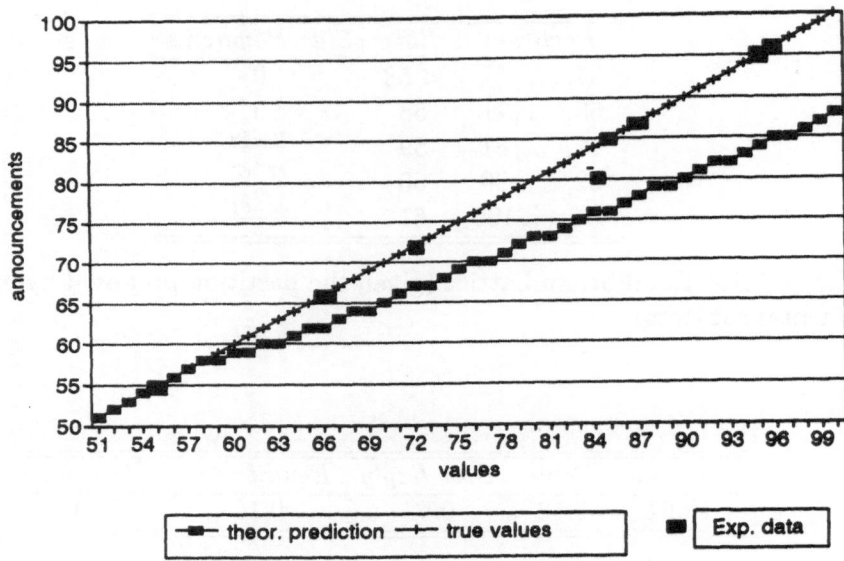

Figure 2.19: First price auction mechanism

2.5.8   First-Price Auction Mechanism

In the last three rounds of Auction 1, players used a first-price auction
mechanism to select the winner of the legitimate auction. As we already observed in Subsection 2.4.2, the mechanism used by experimental
subjects differ from the mechanism suggested by McAfee and McMillan (1992). Experimental subjects splitted equally the difference of the
highest bid minus the reserve price. Details about the theoretical solution of this mechanism (such as used by experimental subjects) are
provided in Section 4.5. Figure 2.19 compares the equilibrium bid for
the prior first-price auction with the bids submitted by experimental
subjects. Note that in all observations but one, experimental subjects
submitted their own value. Note also that, since we use integer numbers,
the equilibrium bid function is not strictly monotone, that is, there is a
positive (but small) probability that this mechanism does not efficiently
allocate the item. The total ex ante expected loss due to this restriction
is just equal to 0.007.

2.5.9   Summing Up

In this section, we have compared experimental data with theoretical
predictions and we have seen that experimental subjects do not play

according to the latter. For the most often used mechanism we have tried to explain experimental subjects' behavior.

Considering the expected payoffs in equilibrium, we can rank the observed mechanisms. The best mechanisms are the first-price auction and the bid-bargain mechanism. They are almost optimal, i.e., in a continuous version of these models, the probability of ties is zero. The second best mechanism is the lattice mechanism. We will see in Section 5.1 that, allowing continuous bids, the inefficiency can completely be eliminated but the expected payoff remains lower than that of an optimal mechanism[35]. Finally, the announcement mechanism gives the lowest expected payoff. Its sub-optimality persists even if we use real numbers.

Considering the incentive to deviate from naïve play in each mechanism, we can say that, for the bid-bargain as well as for the lattice mechanism, there is no high incentive to deviate (see Table 2.12 and 2.16). For the announcement mechanism, on the contrary, the incentive to play the best reply, if the other two players play naïvely is higher (see Table 2.11).

2.6 Other Mechanisms

In Section 2.4, we have described the mechanisms used by experimental subjects. In this section, we want to focus on other mechanisms which have not been used but have been proposed and discussed by them.

As we have already seen in Subsection 2.4.2, experimental subjects who played Auction 1 used the random announcement mechanism for the first 17 rounds. Player 3 realized already in Round 1 that it is possible to cheat when players announce their values in sequence. In order to prevent the players with high values from announcing always as the last one, Player 3 proposed to use a random device to fix the order of the announcements. Still, he was not happy since he recognized the possibility of cheating even in position one and two. In Round 18, Player 1 who until then had denied this possibility by saying that this would be amoral realized that Player 3 was right. He therefore proposed to use a kind of oral auction. Each player had to increase his bid by 10 each time. He did not finish to explain how the mechanism exactly works since Player 3 interrupted him to propose to use the first-price auction mechanism. Players agreed at once on this last proposal. They regretted to have had this idea only at the end of the game.

Although the first-price mechanism was used only once, a sealed-bid

---

[35]This happens since, as we will see, although the player with the highest value obtains the item, he pays $r + \varepsilon$ (with $\varepsilon > 0$) instead of r.

mechanism was proposed in Auction 3 as well. In Round 9, Player 1 suggested to write separately the values on pieces of paper and to turn them simultaneously. His proposal remained ignored.

Another proposed but not used mechanism works as follows. All players who want to obtain the object bid 58, the reserve price, and then, if necessary, the winner is selected by throwing a dice. This mechanism was proposed twice, in Auction 6 and in Auction 11. Experimental subjects did not adopt it because it is not optimal. There is a high probability that the player with the highest value does not obtain the item.

Because of this non-optimality, Player 2 in Auction 11 suggested to use the announcement mechanism. He proposed to avoid the side payments because in each round the probability to be the winner is 1/3 for each player, that is, all players would get on average the same payoff in 20 rounds. Players 1 and 3 did not agree with Player 2. Firstly, he implicitly assumed that they would announce their true values. When it is not the case, there is a high incentive for each player to overannounce his value. Secondly, even though it would be the case that players announce their true values, 20 rounds are not enough in order to make all three players likely to realize almost the same gain at the end of the game. For these reasons, players in Auction 11 opted in favor of the announcement mechanism.

The announcement mechanism without side payments was used in the first round of Auction 9. Player 2 insisted on using a mechanism which could reflect the different positions: who had the highest value should gain all. Player 1 and Player 3 did not agree with him. Player 3 proposed to use the announcement mechanism with side payments splitting equally the highest (supposed) value minus the reserve price. In this way, he argued, one can equally share the risk. A compromise between these two antithetical positions was found in Round 12, when players started to bargain. This allowed to the players with higher values to get more.

Another kind of announcement mechanism was proposed in Auction 8. It consisted of a mechanism in which the player who makes the highest announcement makes side payments but not necessarily splits the surplus equally. Players 2 and 3 rejected this proposal. They preferred to share risk and gains equally.

We conclude this section making two remarks. Firstly, note that the result of the announcement mechanism with side payments is not the same as the result of the announcement mechanism without side payments. This was already understood by Players 1 and 3 in Auction 11. When players play without making side payments, it is a dominant

	End Effect (−)	End Effect (+)
Strategic Behavior (−)	4	2
Strategic Behavior (+)	7	3

Table 2.17: Strategic Behavior and the End Effect

strategy in each position to announce always 100 if $v \geq 58$ and the true value otherwise. That is, the winner is selected always randomly. This mechanism gives the same result of a mechanism in which the players who want to obtain the item bid 58 and the others less than 58. Secondly, note that in most of the auctions risk sharing arguments induce experimental subjects to opt in favor of an equal division of the surplus.

2.7 Strategic Behavior and the End Effect

We have already observed the following two facts. Firstly, there are auctions in which players behave strategically, that is they do not necessarily directly or indirectly announce their true values. Secondly, there are auctions in which cooperation breaks down in the last rounds. This fact is known as the end effect. We now want to investigate which relation there is between strategic behavior and the end effect.

Table 2.17 reports all four possible cases. (+) means that the strategic behavior (or the end effect, respectively) was observed. On the contrary, (−) means that the strategic behavior (or the end effect, respectively) was not observed. To produce this table we have used the following conventions. For the announcement mechanism, we say that we observed strategic behavior in one auction if at least one player cheated in some round. For the bid-bargain mechanism, we say that we always observed strategic behavior. Finally, for the lattice mechanism, we say that we did not observe any strategic behavior[36]. The end effect is (+) for all auctions for which the promised side payments were not made in some rounds, otherwise it is (−)[37].

There are four auctions in which no strategic behavior and no end

---

[36]We do not take the first-price auction mechanism into account in this classification because it was used in just three rounds.

[37]There is one exception: Auction 1. In the last round, Player 3 transferred more than what he promised. He had a bad conscience for having cheated in the previous rounds. We do not classify this auction among those with the end effect.

effect were observed (Auction 7, Auction 9, Auction 10, Auction 14)[38]. Players played always according to the mechanism which they agreed on. They were always sincere. From the analysis of the protocols as well as from the analysis of the written explanations given by experimental subjects, we can conclude that no player of these four groups planned to break the cooperation in the last rounds if he had had the possibility. On the contrary, Player 3 in Auction 9, Round 2 wrote: "I think it is optimal that until the last round, the player with the highest value gets the item and splits equally the difference of his value minus 58."[39]. Nevertheless, they were aware of the advantages to break the agreement in the last rounds. Player 1 in the last round of Auction 10 wrote: "In the last round as well, I stick to the same principle, although it would be worthwhile to consider the possibility to announce a very high value, to get the item and not to make any side payment. But this is in contrast with the continuity as well as, in some way, with moral aspects."[40].

There are seven auctions in which strategic behavior was observed, but no end effect (Auction 1, Auction 4, Auction 6, Auction 8, Auction 12, Auction 13, Auction 15). Some players cheated but they never broke the agreements. One player wrote that he would have broken the cooperation, if he had drawn a high value in the last rounds (Player 1, in Auction 4). The reason why players stick to cooperation until the last round is that they do not want to appear shabby. Player 3, in Auction 8 cheated but in the last round he wrote: "Other people are kind. Me too.". In fact, if a player cheats there is no possibility to discover it. But if he defects, it will be discovered. "Moral aspects" force him to be "correct" until the end of the game.

"It was a surprise; I would have never thought that other players would really cheat"[41]. This wrote Player 2 in the last round of Auction 5 after the other two players admitted they have cheated. Player 3 deviated

---

[38]We classify Auction 9 in this group although players switched for seven rounds to the bid-bargain mechanism. It is interesting to note that players announced their true values in the last round, when they restarted to play according to the announcement mechanism.

[39]In German: "Meiner Meinung nach ist es bis zur 20sten Runde optimal, daß immer der Spieler mit dem höchsten Wiederverkaufswert den Zuschlag erhält und gerecht geteilt wird."

[40]In German: " Auch in der letzten Runde erhalte ich das bisher praktizierte Prinzip aufrecht, obwohl es überlegenswert wäre, gerade in der letzten Runde den übrigen Spielern einen höheren eigenen Wiederverkaufswert zu suggerieren und dann, nachdem man den Zuschlag erhalten hat, keine Gewinnverteilung vorzunehmen. Dem stehen jedoch die Kontinuität sowie in gewisser Weise auch moralische Aspekte entgegen."

[41]In German: Das war eine Überraschung; hätte nicht gedacht, daß die anderen wirklich geschummelt haben. Erstaunliche Welt.

in Round 18. He did not make any side payment. Already in Round 2, Player 3 wrote that he would cooperate until Round 17 or 18. In the other two auctions in which players cheated and defected at the end, there were not such reactions. Other players have planned to do the same. In Auction 11, there was also an anticipated end effect in Round 7. But they started again to cooperate, although they were very cautious.

Only in two cases, we observed players announcing sincerely their values and to defect at the end (Auction 2, Round 18 and Auction 16, Round 19). For Auction 2, we have no evidence that players realized that they could cheat. Even after Player 1's deviation, they did not ask themselves and/or other players whether some other player had cheated. We asked Player 1 who played Auction 16 and who deviated in Round 19, whether she had realized that there was a possibility to cheat. She answered affirmatively, but she realized it only in Round 12. "It would have been too difficult for me to cheat", she said. "Other players could realize that you are hesitating if you are not fast enough".

Even if the end effect appeared only in five auctions, we have enough evidence to say that it could have appeared more often, if moral aspects had not refrained experimental subjects from deviating in the last rounds. On the contrary, strategic behavior manifested itself in many auctions since it was not detectable. This is a good way to "maximize" one's own payoff, to appear kind and to guard oneself against eventual deviations at the end of the game.

Chapter 3

A Descriptive Model

In this chapter, we present a descriptive model which characterizes experimental subjects' behavior when they play the announcement mechanism. The observation of experimental data and, particularly, some regularities in the deviation of data on average from theoretical predictions suggest that there is a possible explanation of aggregated data. We have observed that players are divided into two groups: players who cheat and players who do not cheat (see Subsection 2.5.2). Moreover, the average values from which players start to cheat are very close for the three different positions (81.7, 81.2 and 80.5). We will say that a value is low when it is strictly lower than 81, and high when it is equal to or higher than 81. As we have seen in Subsection 2.5.2, players who cheat do not do so according to the theoretical prediction. There, we have proposed a simple model that describes how players who cheat announce their values. Our aim, in this section, is to present a more general model, which assigns the conditional probability that a player cheats when he announces in the different positions, given his value and the others'announcements. To estimate these probabilities, we use a logit model. After having estimated the probability that a player cheats when he announces in one of the three positions (Section 3.1), we show how the descriptive model works (Section 3.2). We simulate players' behavior in our descriptive model, and we compare the simulated players' behavior with experimental subjects behavior.

3.1 Conditional Probabilities

We start by introducing some notation. Let Y_i^1 and Y_i^2 be the events "player who announces in position i cheats" and "player who announces in position i says his own value", respectively. Let X_i^1 and X_i^2 be the events "player who announces in position i has a high value" and "player who announces in position i has a low value", respectively. Let Z_i^1 and Z_i^2 be the events "player who announces in position i has heard a value which is strictly lower than his own" and "player who announces in position i, $(i = 2,3)$ has heard a value which is equal to or higher than his own value", respectively. Finally, let define the following conditional probabilities

$$P_1^j = P[Y_1^1 \mid X_1^j]$$
$$P_i^{j,k} = P[Y_i^1 \mid X_i^j, Z_i^k]$$

where $j, k = 1, 2$ and $i = 2, 3$. P_1^j is the probability that player in position one cheats, given his value. $P_i^{j,k}$ is the probability that player in position i $(i = 2, 3)$ cheats, given his value and given what he has heard. With self-explanatory notation, we can define

$$1 - P_1^j = P[Y_1^2 \mid X_1^j]$$

$$1 - P_i^{j,k} = P[Y_i^2 \mid X_i^j, Z_i^k]$$

Our next step is the estimation of P_1^j and $P_i^{j,k}$.

Consider the following equation:

$$\ln \frac{P_1^j}{1 - P_1^j} = \gamma_1^0 + \gamma_1^j \tag{3.1}$$

with $j = 1, 2$. The first term of Equation 3.1 is called *logit* of cheating of the player who announces as the first one[1]. It is a probability transformation: this function is equal to $-\infty$ for $P_1^j = 0$, it is equal to 0 for $P_1^j = 0.5$, and is equal to ∞ for $P_1^j = 1$. The second term of Equation 3.1 consists of a constant term, γ_1^0, which does not depend on the value that Player 1 has drawn and of variable, γ_1^j, which depends on the value that Player 1 has drawn. Without loss of generality, we can assume that $\gamma_1^2 = 0$, that is γ_1^0 represents the logit transformation in the case that the player who announces first cheats whenever he has a low value. The first row of Table 3.1 shows the number of cases divided in the two independent variables. The second one shows the number of cases in which the player in position 1 cheats, and the last one shows the relative frequencies of this event. Replacing the probabilities of the first term of Equation 3.1 with the relative frequencies of cheating, and adding error terms ε, one gets

$$\begin{bmatrix} \ln(11/83) \\ \ln(2/143) \end{bmatrix} = \begin{bmatrix} 1 & 1 \\ 1 & 0 \end{bmatrix} \cdot \begin{bmatrix} \gamma_1^0 \\ \gamma_1^1 \end{bmatrix} + \begin{bmatrix} \varepsilon_1 \\ \varepsilon_2 \end{bmatrix} \tag{3.2}$$

---

[1] The adjective *logit* is due to the relation of this transformation with the logistic function (see Theil (1971)).

	X_1^1	X_1^2	Total
Number of cases	94	145	239
Number of cases P1 cheats	11	2	13
Relative Frequency	0.1170	0.0139	0.0544

Table 3.1: Data for the player who announces in position 1

$\hat{P}_1^1$	$\hat{P}_1^2$
0.1171	0.014
(0.103)	(0.507)

Table 3.2: Estimations

This is equivalent to the well known econometric relation $\mathbf{Y} = \mathbf{X}\beta + \varepsilon$. One assumes that relative frequencies are based on samples independently drawn from binomial populations. That implies that the error terms ε are independent as well. Table 3.2 shows estimated values of P_1^1, P_1^2 and estimators of the standard deviation of error terms in brackets. Comparing these values with the relative frequencies, one ascertains the goodness of fit. Following the same procedure, we estimate now $P_i^{j,k}$, $j, k = 1, 2$, and $i = 2, 3$. Since we now have one independent variable more, namely Z_i, we have to take it into account. Equation 3.3 is the logit transformation of $P_i^{j,k}$

$$\ln \frac{P_i^{j,k}}{1 - P_i^{j,k}} = \gamma_i^0 + \gamma_i^{j,k} + \delta_i^{j,k} \tag{3.3}$$

The second term of Equation 3.3 consists of a constant term, γ_i^0, and of two variables, $\gamma_i^{j,k}, \delta_i^{j,k}$. Without loss of generality, we can assume that $\gamma_i^{1,2} = \gamma_i^{2,2} = \delta_i^{1,2} = \delta_i^{2,2} = 0$ in the cases $(X_i = X_i^1, Z_i = Z_i^2)$, $(X_i = X_i^2, Z_i = Z_i^2)$. That is, the logit transformation in Equation 3.3 is equal to γ_i^0 when Player i has heard a value which is equal to or higher than his own value. In the case $(X_i = X_i^2, Z_i = Z_i^1)$, we define the logit transformation to be equal to $\gamma_i^0 + \delta_i^{2,1}$, that is $\gamma_i^{2,1} = 0$. Finally, in the case $(X_i = X_i^1, Z_i = Z_i^1)$, we define the logit transformation to be equal to $\gamma_i^0 + \gamma_i^{1,1} + \delta_i^{1,1}$, $i = 2, 3$. Table 3.3 shows the data for the player who

	(X_2^1, Z_2^1)	(X_2^2, Z_2^1)	(X_2^1, Z_2^2)	(X_2^2, Z_2^2)	Total
Number of cases	65	46	19	109	239
Number of cases P2 cheats	18	2	0	1	21
Relative Frequency	0.28	0.043	0	0.0091	0.088

Table 3.3: Data for the player in position 2

$\hat{P}_2^{11}$	$\hat{P}_2^{21}$	$\hat{P}_2^{12}$	$\hat{P}_2^{22}$
0.277	0.043	0.022	0.022
(0.076)	(0.52)	(1.06)	(1.009)

Table 3.4: Estimated probabilities for the player in position 2

announces as the second one. We repeat the same procedure already used for the first player. Note that since the relative frequency of the event (X_2^1, Z_2^2) is zero, and since the logarithm is not defined at zero, we have to assume that there is a very low (but positive) frequency that Player 2 cheats when he has high values and he has heard a value which is equal to or higher than his own. Again, replacing the probabilities of the first term of Equation 3.3 with the relative frequencies of cheating obtained from Table 3.3, and adding error terms ε, we get

$$\begin{bmatrix} \ln(18/47) \\ \ln(2/44) \\ \ln(1/18) \\ \ln(1/108) \end{bmatrix} = \begin{bmatrix} 1 & 1 & 1 & 0 \\ 1 & 0 & 0 & 1 \\ 1 & 0 & 0 & 0 \\ 1 & 0 & 0 & 0 \end{bmatrix} \cdot \begin{bmatrix} \gamma_2^0 \\ \gamma_2^{1,1} \\ \delta_2^{1,1} \\ \delta_2^{2,1} \end{bmatrix} + \begin{bmatrix} \varepsilon_{11} \\ \varepsilon_{21} \\ \varepsilon_{12} \\ \varepsilon_{22} \end{bmatrix}$$

Table 3.4 shows estimated values of $P_2^{j,k}$ and estimators of the standard deviation of error terms in brackets. Finally, we execute the same analysis for the player who announces as last one. Tables 3.5 and 3.6 show the observed data and the estimated values, respectively.

To estimate $P_i^{j,k}$, we used the method of the weighted OLS. We have assumed that the error terms ε are independent, identically distributed, and drawn from a binomial population. One can show that expected value and variance of $\varepsilon_{j,k}$ are equal to 0 and $1/n_{j,k} f_{j,k}(1 - f_{j,k})$, respectively where $n_{j,k}$ is the number of cases (in the first row) and $f_{j,k}$ is the relative frequency that a player cheats. We weighted our relation with the squared root of the reciprocal of $n_{j,k} f_{j,k}(1 - f_{j,k})$.

	(X_3^1, Z_3^1)	(X_3^2, Z_3^1)	(X_3^1, Z_3^2)	(X_3^2, Z_3^2)	Total
Number of cases	54	23	28	134	239
Number of cases P3 cheats	16	4	1	1	22
Relative Frequency	0.314	0.1̄8̄	0.04	0.00813	0.1

Table 3.5: Data for the player in position 3

$\hat{P}_3^{11}$	$\hat{P}_3^{21}$	$\hat{P}_3^{12}$	$\hat{P}_3^{22}$
0.295	0.173	0.016	0.016
(0.09)	(0.3)	(1.04)	(1.007)

Table 3.6: Estimated probabilities for the player in position 3

3.2 The Model

We are now ready to introduce the descriptive model. This model gene-
ralizes the model presented in Subsection 2.5.2. There, we have proposed
a simple linear model that describes how players who cheat make their
announcements. Here, we describe how and when players cheat in one of
the three positions.

Players use the announcement mechanism. This mechanism works as
follows: the values are drawn from the following distribution

$$p(v_i, v_j, v_k) = \begin{cases} \frac{1}{50^3 - 7^3} & \text{if } \max\{v_1, v_2, v_3\} \geq 58 \\ 0 & \text{otherwise} \end{cases}$$

with $v_i, v_j, v_k \in \{51, \ldots, 100\}$. Each player has to make an announcement
$\hat{v} \in \{51, \ldots, 100\}$. The player who makes the highest announcement,
$\hat{v}_{(1)} = \max\{\hat{v}_1, \hat{v}_2, \hat{v}_3\}$, where $\hat{v}_i$ is the announcement of the player who
announces in position i, pays the reserve price $r = 58$ to the auctioneer,
and pays each losing bidder an equal share of the difference between his
announcement and the reserve price[2]. The side payments may at most
be equal to the realized payoff in each round, that is $(2/3)(\hat{v}_i - r) \leq$
$\max\{0, v_i - r\}$, for each $i = 1, 2, 3$. If two or all players announce $\hat{v}_{(1)}$ a fair
dice is thrown to select the winner. If no player makes an announcement

---

[2]Note that we are disregarding the units problem.

equal to or higher than 58, the item is not bought. We assume that, given that players agree on this mechanism, no player will defect.

This model wants to describe players' announcements in the different positions. Using the notation introduced in Section 3.1, we have that the player who announces as the first one announces $\hat{v}_1 = \hat{v}_1^*(v_1)$ as in Equation 3.4.

$$\hat{v}_1^*(v_1) = \begin{cases} v_1 & 1 - P_1^j \\ \beta_1 \, v_1 & P_1^j \end{cases} \tag{3.4}$$

That is, having drawn v_1, Player 1 announces his own value with probability $1 - P_1^j$ and $\beta_1 \, v_1$ with probability P_1^j. If v_1 is high, that is if $v_1 \geq 81$ then we have $P_1^j = P_1^1$, otherwise we have $P_1^j = P_1^2$.

Player 2 hears Player 1's announcement $\hat{v}_1$. If his value is equal to or lower than Player 1's announcement then he announces according to Equation 3.5.

$$\hat{v}_{*2}(v_2) = \begin{cases} v_2 & 1 - P_2^{j,2} \\ \beta_2 \, v_2 & P_2^{j,2} \end{cases} \tag{3.5}$$

Equation 3.5 says that having drawn v_2 and having heard $\hat{v}_1$, Player 2 announces his own value with probability $1 - P_2^{j,2}$ and $\beta_2 \, v_2$ with probability $P_2^{j,2}$. Otherwise, if his value is higher than Player 1's announcement, he announces the maximum of $\hat{v}_2^*$, defined in Equation 3.6, and $\hat{v}_1 + 1$. Equation 3.6 says that having drawn v_2 and having heard $\hat{v}_1$, Player 2 announces his own value with probability $1 - P_2^{j,1}$ and $\beta_2 \, v_2$ with probability $P_2^{j,1}$.

$$\hat{v}_2^*(v_2) = \begin{cases} v_2 & 1 - P_2^{j,1} \\ \beta_2 \, v_2 & P_2^{j,1} \end{cases} \tag{3.6}$$

Note that $P_2^{j,1}$ as well as $P_2^{j,2}$ depend on v_2. If $v_2 \geq 81$ then we have $j = 1$, otherwise, we have $j = 2$.

Finally, the player who announces as the third one has heard $\hat{v}_1$ and $\hat{v}_2$, i.e., Player 1's and Player 2's announcement. If his value is equal to or lower than $\max\{\hat{v}_1, \hat{v}_2\}$ then he announces $\hat{v}_3$ as defined in Equation 3.7.

$$\hat{v}_{*3}(v_3) = \begin{cases} v_3 & 1 - P_3^{j,2} \\ \beta_{31} \, v_3 + \beta_{32} \, v_m & P_3^{j,2} \end{cases} \tag{3.7}$$

where $j, k = 1, 2$, $v_m = \max\{\hat{v}_1, \hat{v}_2\}$. This equation says that having drawn v_3 and having heard $\hat{v}_1$ and $\hat{v}_2$, Player 3 announces his own value with probability $1 - P_3^{j,2}$ and $\beta_{31} \, v_3 + \beta_{32} \, v_m$ with probability $P_3^{j,2}$.

If Player 3's value is higher than $\max\{\hat{v}_2, \hat{v}_3\} \geq 58$, he announces $\max\{\hat{v}_3^*, \{\max\{\hat{v}_1, \hat{v}_2\} + 1\}\}$, with $\hat{v}_3^*$ defined in Equation 3.8.

$$\hat{v}_3^*(v_3) = \begin{cases} v_3 & 1 - P_3^{j,1} \\ \beta_{31} \, v_3 + \beta_{32} \, v_m & P_3^{j,1} \end{cases} \tag{3.8}$$

where $j, k = 1, 2$, $v_m = \max\{\hat{v}_1, \hat{v}_2\}$.

This equation says that having drawn v_3 and having heard $\hat{v}_1$ and $\hat{v}_2$, Player 3 announces his own value with probability $1 - P_3^{j,1}$ and $\beta_{31} v_3 + \beta_{32} v_m$ with probability $P_3^{j,1}$. If $v_3 \geq 58 > \max\{\hat{v}_2, \hat{v}_3\}$, Player 3 announces $\max\{\hat{v}_3^*, 58\}$, with $\hat{v}_3^*$ as defined in Equation 3.8. Note that the ex post probability to cheat depends on the observation of a low or high value. It does not matter if the observed value is 85 or 94. It matters only if the value is high or low. Making this assumption, we can easily estimate these probabilities. Moreover, we neither need to know the ex ante probability that a player cheats in a position.

The price we have to pay for using the logit model, which we use to estimate these probabilities, is not too high. Firstly, we have to assign a positive probability to the event that a player (in position two as well as in position three) will cheat having heard an announcement higher than his own value. We can justify this fact saying that there is a low probability that a player forgets his exact value[3]. Secondly, using the logit model we have to assume that relative frequencies are based on samples independently drawn from binomial populations. This assumption appears realistic. One can imagine that players' population is binomial, there are players who are strategically sophisticated and players who are naïve. The problem is that players' behavior is also influenced by the group composition.

To model how players announce when they cheat, we have chosen a very simple structure. This fits very well our data as we have already seen in Subsection 2.5.2.

The last step is the simulation of players' behavior of our descriptive model and the comparison of simulated data with experimental data. We simulated this model for 150000 rounds. In each round, each player draws a value from a uniform discrete distribution on $\{51, \ldots, 100\}$. If all three values are lower than 58 then players draw new values. We use the estimated values of the probabilities obtained in Section 3.1 and estimated values of betas obtained in Subsection 2.5.2[4].

Figures 3.1, 3.2 and 3.3 show the results of the simulation and compare aggregated experimental results with the simulated results. We want to stress that we do not assign any predictive meaning to this simulation. This simulation describes players' behavior if this game is played by a large number of (unexperienced) groups.

---

[3]Actually, this happened for a player who announced as the third one (see Table 3.5).

[4]We use the estimated values of β_{31} and β_{32} given in Table 2.9.

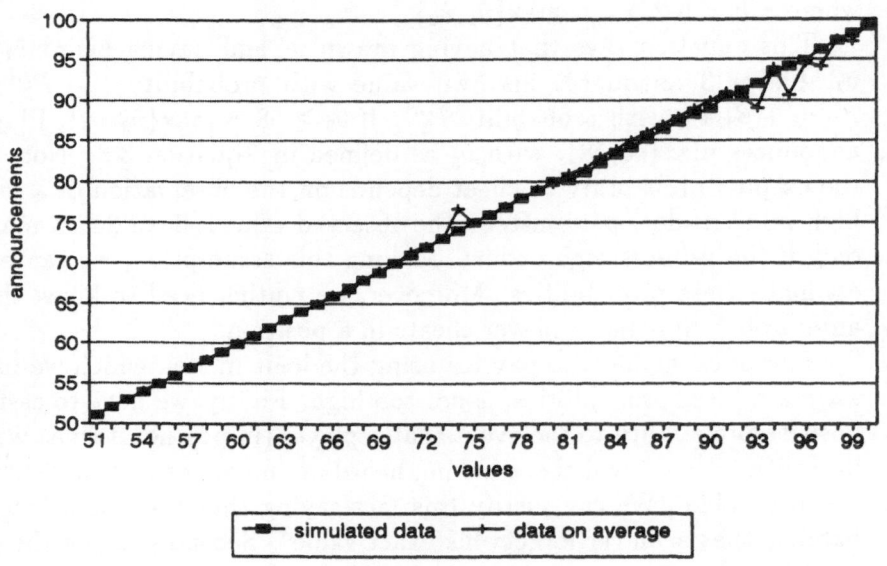

Figure 3.1: Simulated data (Player 1)

Figure 3.2: Simulated data (Player 2)

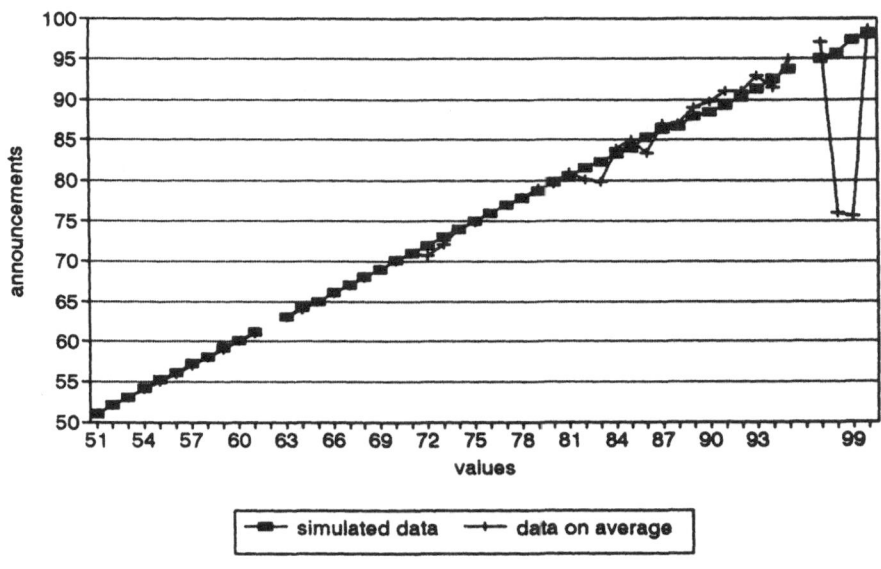

Figure 3.3: Simulated data (Player 3)

Chapter 4

Mechanisms of Collusion

In this chapter, we analyze the mechanisms presented in Chapter 2. We will start showing how the probability distribution of the values in our experimental set-up looks like (Section 4.1). Section 4.2 analyzes a particular equilibrium of the announcement mechanism, Section 4.3 the equilibria of the bid-bargain mechanism. In Subsection 4.3.1, we investigate the optimality of the bid-bargain mechanism and compare it with the announcement mechanism. In Section 4.4, we show an algorithm to calculate the equilibria of the lattice mechanism. The optimality of this mechanism is investigated in Subsection 4.4.1. As we will see, the lattice mechanism can give very bad expected payoffs depending on the assigned partition of the values. In Subsection 4.4.2, we show the partition with four intervals which guarantees the highest expected payoff for the set of values of our experiment. The first-price auction mechanism is analyzed in Section 4.5. Section 4.6 sums up the most important results.

4.1 The Distribution Function

We show here how the conditional distribution function of three discrete random variables looks like, when each of them is drawn with a bernoullian sampling procedure from the same support $\{51, \ldots, 100\}$, and the maximum has to be equal to or higher than 58.

Hereafter, random variables are always typed in bold style. Let us define the probability function $p(v_1, v_2, v_3 \mid \mathbf{v}_{(1)} > 57)$, that is the probability that $\mathbf{v_1} = v_1$, $\mathbf{v_2} = v_2$ and $\mathbf{v_3} = v_3$, given that $\mathbf{v}_{(1)} > 57$, where $\mathbf{v}_{(i)}$ is the ith order statistic of $\mathbf{v} = (\mathbf{v_1}, \mathbf{v_2}, \mathbf{v_3})$, so that $\mathbf{v}_{(1)} \equiv \max\{\mathbf{v_1}, \mathbf{v_2}, \mathbf{v_3}\} \geq \mathbf{v}_{(2)} \geq \mathbf{v}_{(3)}$. This is equal to $p(v_1, v_2, v_3, v_{(1)} > 57)/p(v_{(1)} > 57)$. With these definitions, we are able to define the conditional distribution function of the random variables $\mathbf{v_1}$, $\mathbf{v_2}$, $\mathbf{v_3}$ given that $v_{(1)} > 57$.

$$F_{v_1,v_2,v_3}(x_1, x_2, x_3 \mid v_{(1)} > 57) = \sum_{v_1 \leq x_1, v_2 \leq x_2, v_3 \leq x_3} p(v_1, v_2, v_3 \mid v_{(1)} > 57)$$

Note that $p(v_1, v_2, v_3 \mid v_{(1)} > 57)$, the probability function of the random variables v_1, v_2, v_3 given that $v_{(1)} > 57$, is uniform. Having this probability function to hand, we can calculate all we need. For example, we can calculate players' total ex ante expected payoff when they cooperate[1]. This is equal to $E(v_{(1)} \mid v_{(1)} > 57) - 58$ [2]. Using the conditional distribution function $F_{v_1,v_2,v_3}(\cdot \mid v_{(1)} > 57)$, one can derive the probability function of the maximum of the random variables v_1, v_2, v_3 given that $v_{(1)} > 57$. Calculating it, one finds out that $E(v_{(1)} \mid v_{(1)} > 57)$ is approximately equal to 88.08. That is the total expected payoff is approximately equal to 30.08. We can also calculate players' ex ante payoff when they do not cooperate[3]. This value is approximately equal to 4.20.

It is of interest in our set-up to consider the ex post probability function of two random variables given that one random variable has been observed, and the ex post probability function of one random variable given that two random variables have been observed.

$$p(v_i, v_j \mid v_k = y) = \begin{cases} \frac{1}{50^2} & \text{if } y \geq 58 \\ \frac{1}{50^2 - 7^2} & \text{if } y < 58 \text{ and } v_{(1)} \geq 58 \\ 0 & \text{otherwise} \end{cases} \quad (4.1)$$

$$p(v_i \mid v_j = z, v_k = y) = \begin{cases} \frac{1}{50} & \text{if } \max(z, y) \geq 58 \\ \frac{1}{50 - 7} & \text{if } \max(z, y) < 58 \text{ and } v_i \geq 58 \\ 0 & \text{otherwise} \end{cases} \quad (4.2)$$

---

[1] We use cooperating meaning what we have meant in Chapter 2, i.e., one player submits a serious bid and the others phony bids.

[2] Actually, this is players' total ex ante expected payoff when they use an optimal mechanism, that is, when they use a mechanism which selects as winner the player with the highest value.

[3] Calculating players equilibrium bids when they do not cooperate, one has to take into account that players' values are not independent. Consider the problem that Player i has to solve, after that he has observed his value v_i. If $v_i \leq 57$ then it is optimal to bid $b(v) \leq 57$. Otherwise, he could incur in a loss. Therefore, we can restrict ourselves to the case $v_i \geq 58$. Since the relevant distribution to him is the distribution of $y_{(1)} = \max\{v_j, v_k\}$, with $j, k \neq i$, given that $v_{(1)} > 57$, $F_{y_{(1)}}(\cdot \mid v_{(1)} > 57)$, and since this conditional distribution, in our case, does not depend on v_i, we have the following equilibrium bid function (for the continuous version) $b^*(v) = \frac{2}{3}v + \frac{50}{3} + \frac{(58-50)^3}{3 \cdot (v-50)^2}$. This expression is equal to the well known expression for independent random variables.

Equations 4.1 and 4.2 have been calculated by Bayes' rule, using the fact that $p(v_i, v_j, v_k)$ is equal to $\frac{1}{50^3 - 7^3}$ if $v_{(1)} \geq 58$ and it is equal to zero otherwise. Note that since the maximum of the three values has to be equal to or higher than 58, the probability to draw a value equal to or higher than 58, given that the maximum of the values that have been already drawn is lower than 58, is higher than the probability to draw the same value, given that the maximum of the values that have been already drawn is equal to or higher than 58. Now, we are ready to start to analyze each single mechanism.

4.2 Announcement Mechanism

In this section we examine the solution of the announcement mechanism. Hereafter, writing $[x]$, we mean the integer part of x. As we have already seen in Chapter 2, this mechanism works as follows: players draw their values from the following distribution

$$p(v_i, v_j, v_k) = \begin{cases} \frac{1}{50^3 - 7^3} & \text{if } v_{(1)} \geq 58 \\ 0 & \text{otherwise} \end{cases}$$

with $v_i, v_j, v_k \in \{51, \dots, 100\}$. Each player has to make an announcement $\hat{v} \in \{51, \dots, 100\}$. The player who makes the highest announcement, $\hat{v}_{(1)} = \max\{\hat{v}_1, \hat{v}_2, \hat{v}_3\}$, where $\hat{v}_i$ is the announcement of the player who announces in position i, pays the reserve price $r = 58$ to the auctioneer, and pays each losing bidder $[(\hat{v}_{(1)} - r)/3]$. If more players make the highest announcement, a fair dice is thrown to choose the winner. The side payments may at most be equal to the realized payoff in each round, that is $(2/3)(\hat{v}_i - r) \leq \max\{0, v_i - r\}$, for each $i = 1, 2, 3$. Since, as we have already seen in Chapter 2, when $3 \cdot ((\hat{v}_{(1)} - r)/3 - [(\hat{v}_{(1)} - r)/3]) > 0$, that is, when $\hat{v}_{(1)} - r$ is not a multiple of 3, the winner obtains this difference in most of the auctions, we assume this in our model. If no player makes an announcement equal to or higher than 58, the item is not bought. We assume that, given that players agree on this mechanism, no player will defect[4].

As we have already seen in Chapter 2, this mechanism is an idealization of the true mechanism used by experimental subjects. However, this idealization is not very different from the mechanism used by experimental subjects.

---

[4]This assumption is necessary since experimental subjects may not sign contracts in our experiment.

In this section, whenever it is not confusing, we will write Player i instead of "the player who announces in position i". We want to calculate the best announcement for Player $i, (i = 1, 2, 3)$. We will proceed as follows. We start analyzing Player 3's best announcement, given Player 1's and Player 2's announcement. Then, given Player 1's announcement, and anticipating Player 3's behavior, we find Player 2's best announcement. Finally, anticipating Player 2's and Player 3's behavior, we characterize Player 1's best announcement. We restrict ourselves to equilibria in pure strategies. In Chapter 2, we have already given the intuition of the solution. Here, we want to be more precise and give all details.

To simplify our notation, in what follows, we do not explicitly write that each probability is conditioned to the event that $\mathbf{v}_{(1)} \geq r$.

Let $\varphi = (\varphi_1(v_1), \varphi_2(\hat{v}_1, v_2), \varphi_3(\hat{v}_1, \hat{v}_2, v_3))$ be a strategy profile. When $v_2 \leq 57$ and $\hat{v}_1 \leq 59$, Player 2 can neither conclude that $v_1 \leq 57$ nor that $v_1 \geq 58$ (see the proof of Proposition 1), that is, when $v_2 \leq 57$ and $\hat{v}_1 \leq 59$, Player 2's local payoff function, i.e. Player 2's expected payoff after that he has heard $\hat{v}_1$, depends on Player 1's type. To calculate Player 2's local payoff function, we introduce beliefs about Player 1's type. Let be

$$\sigma(\hat{v}_1 \mid v_1) = \begin{cases} 1 & \text{if } \varphi_1(v_1) = \hat{v}_1 \\ 0 & \text{otherwise} \end{cases}$$

and $\mu(v_1 \mid \hat{v}_1)$ the ex post probability that Player 1's type is v_1, given that he has announced $\hat{v}_1$. The equilibrium we are looking for is a *perfect Bayesian equilibrium* (Fudenberg and Tirole (1991)) in pure strategies. However, the definition we are giving is not the general one. We give a special definition adapted to the game examined here. The reason is that, as we have already said, beliefs are important only for Player 2. Let $h_i(v_i, \hat{v}_1, \hat{v}_2, \hat{v}_3)$ be Player i's payoff, if his value is v_i and $\hat{v}_1, \hat{v}_2, \hat{v}_3$ are players' announcements. That is, $h_i(v_i, \hat{v}_1, \hat{v}_2, \hat{v}_3)$ is Player i's payoff in the extensive game. Let π_i be Player i's local payoff at (φ, μ) which is defined as follows

$$\pi_3(v_3, \hat{v}_1, \hat{v}_2, \hat{v}_3) =$$
$$h_3(v_3, \hat{v}_1, \hat{v}_2, \hat{v}_3)$$
$$\pi_2(v_3, \hat{v}_1, \hat{v}_2, \varphi_3, \mu) =$$
$$\sum_{v_1=51}^{100} \mu(v_1 \mid \hat{v}_1) \sum_{v_1=51}^{100} p(v_3 \mid v_1, v_2) h_2(v_2, \hat{v}_1, \hat{v}_2, \varphi_3(\hat{v}_1, \hat{v}_2, v_3))$$
$$\pi_1(v_1, \hat{v}_1, \varphi_2, \varphi_3) =$$
$$\sum_{v_2=51}^{100} \sum_{v_3=51}^{100} p(v_2, v_3 \mid v_1) h_1(v_1, \hat{v}_1, \varphi_2(\hat{v}_1, v_2), \varphi_3(\hat{v}_1, \hat{v}_2, v_3))$$

Definition 3 *A perfect Bayesian equilibrium of the game defined above is a strategy profile φ^* and posterior beliefs $\mu(\cdot \mid \hat{v}_1)$ of Player 2 such that*

P1 $\forall v_1$ we have

$$\hat{v}_1^* = \varphi_1^*(v_1) = \arg\max_{\hat{v}_1} \pi_1(v_1, \hat{v}_1, \varphi_2^*, \varphi_3^*)$$

P2 $\forall v_2$ and $\hat{v}_1^*$ we have

$$\hat{v}_2^* = \varphi_2^*(\hat{v}_1^*, v_2) = \arg\max_{\hat{v}_2} \pi_2(v_2, \hat{v}_1^*, \hat{v}_2, \varphi_3^*, \mu)$$

P3 $\forall v_3$, $\hat{v}_1^*$ and $\hat{v}_2^*$ we have

$$\hat{v}_3^* = \varphi_3^*(\hat{v}_1^*, \hat{v}_2^*, v_3) = \arg\max_{\hat{v}_3} \pi_3(v_3, \hat{v}_1^*, \hat{v}_2^*, \hat{v}_3)$$

B

$$\mu(v_1 \mid \hat{v}_1) = \frac{p(v_1 \mid v_2) \cdot \sigma(\hat{v}_1 \mid v_1)}{\sum_{v_1'=51}^{100} p(v_1' \mid v_2) \cdot \sigma(\hat{v}_1 \mid v_1')}$$

if $\sum_{v_1'=51}^{100} p(v_1' \mid v_2) \cdot \sigma(\hat{v}_1 \mid v_1') > 0$ and $\mu(\cdot \mid \hat{v}_1)$ is any other probability distribution on $\{51, \ldots, 100\}$ if

$$\sum_{v_1'=51}^{100} p(v_1' \mid v_2) \cdot \sigma(\hat{v}_1 \mid v_1') = 0$$

P1 until P3 are the perfection conditions. P1 says that Player 1 takes into account the effect of $\hat{v}_1$ on Player 2's action; P2 states that Player 2 reacts optimally to Player 1's action given his posterior beliefs about v_1; P3, finally, states that Player 3 reacts optimally to Player 1's and Player 2's actions. B corresponds to the Bayes' rule. Note that if $\hat{v}_1$ is not part of Player 1's optimal strategy for some type v_1, observing $\hat{v}_1$ is a probability-0 event. In this case, any posterior beliefs $\mu(\cdot \mid \hat{v}_1)$ are admissible. Note the link between strategies and beliefs: the beliefs are consistent with the strategies, which are optimal given the beliefs. As we will see below, Player 2 will need $\mu(\cdot \mid \hat{v}_1)$ only in the case in which $\hat{v}_1 \leq 59$ and $v_2 \leq 57$.

Since, as we will see, the best announcements are not unique, we restrict the set of equilibria by introducing the two following properties[5].

---

[5]I am in debt to professor Selten who suggested to introduce these two properties to obtain only one equilibrium.

MUD *Minimality of unprofitable dishonesty.*

   If the cardinality of $\arg\max_{\hat{v}_i} \pi_i(v_i, \hat{v}_i, \ldots)$ is strictly higher than one, then an optimal announcement of Player i, $\hat{v}_i^*$, is such that $\hat{v}_i^* \in \arg\min_{\tilde{v}_i}\{|\ \tilde{v}_i - v_i\ |\}$ with $\tilde{v}_i \in \arg\max_{\hat{v}_i} \pi_i(v_i, \hat{v}_i, \ldots)$.

NSA *Non-strategic attribution of deviations.*

   If
$$\sum_{v_1'=51}^{100} p(v_1'\mid v_2)\cdot \sigma(\hat{v}_1\mid v_1') = 0$$

   then
$$\mu(v_1\mid \hat{v}_1) = \begin{cases} 1 & \text{if}\ \ v_1 = \hat{v}_1 \\ 0 & \text{otherwise} \end{cases}$$

MUD says that when Player i has more than one optimal announcement, he will always announce the optimal announcement which minimize the absolute value of the difference $(\tilde{v}_i - v_i)$, with $\tilde{v}_i \in \arg\max_{\hat{v}_i} \pi_i(v_i, \hat{v}_i, \ldots)$.

This implies that if v_i is an optimal announcement, Player i will announce v_i. That is, when cheating is profitable, players prefer to announce the optimal value closest to their true value. Note that if the cardinality of $\arg\min_{\tilde{v}_i}\{|\ \tilde{v}_i - v_i\ |\}$ is equal to two, Player i will be indifferent between announcing one of the two values which minimize $\arg\min_{\tilde{v}_i}\{|\ \tilde{v}_i - v_i\ |\}$. NSA says that when Player 2 hears an announcement which is a probability-0 event, he will believe that Player 1 announces his true value.

These two properties are not completely ad hoc. Their introduction is justified by the experimental observations. As we have already seen in Chapter 2, experimental subjects announce their true value in most of the rounds. Moreover, we have seen in the descriptive model introduced in Chapter 3 that players overbid the theoretical prediction when they cheat. Note also that these two properties justify each other. In fact, if we assume that players want to minimize the amount of unprofitable dishonesty, it seems to be natural to assume that when Player 2 hears an announcement which is a probability-0 event, he believes that Player 1 wants to minimize the dishonesty, even when this would be profitable for him. On the contrary, assuming NSA, it seems to be be natural to assume that players minimize the amount of cheating when this is costless.

We look for perfect Bayesian equilibria which satisfy the MUD and NSA properties. The following result says that there is only one perfect Bayesian equilibrium which satisfies the MUD and NSA properties.

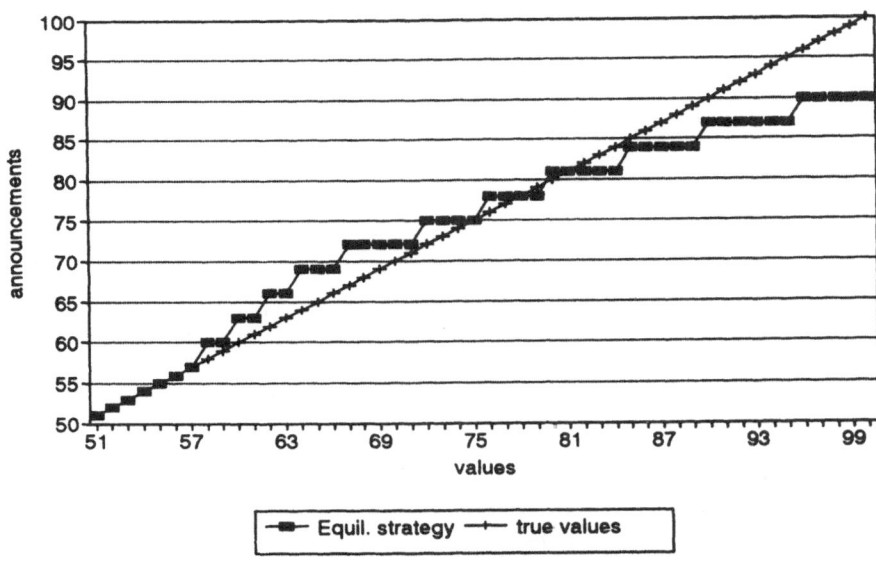

Figure 4.1: Player 1's equilibrium strategy

Proposition 1 The following pair $(\varphi^*, \mu(\cdot \mid \hat{v}_1))$ is the only perfect Bayesian equilibrium which has the MUD and NSA properties.

1. $\mu(\cdot \mid \hat{v}_1))$ is calculated according to Bayes' rule if $\sum_{v_1'=51}^{100} p(v_1' \mid v_2) \cdot \sigma(\hat{v}_1 \mid v_1') > 0$, and it is defined according to NSA otherwise.

2. Player 1's equilibrium strategy, φ_1^*, is defined as in Figure 4.1.

3. Player 2's equilibrium strategy, φ_2^*, is defined as follows:

   If $v_2 < 58$ then

$$\varphi_2^*(\hat{v}_1^*, v_2) = \begin{cases} v_2 & \text{if } \hat{v}_1^* \geq 60 \\ 60 & \text{otherwise} \end{cases} \tag{4.3}$$

   If $v_2 \geq 58$ then φ_2^* is the function tabulated in Appendix E.

4. Player 3's equilibrium strategy, φ_3^*, is defined by Expressions 4.4 until 4.8 as follows:

   Let $\hat{y}_{(1)}$ be the highest announcement Player 3 has heard, that is, $\hat{y}_{(1)} = \max\{\hat{v}_1^*, \hat{v}_2^*\}$. Let be $\hat{y}_{(1)} \geq 58$. For $\hat{y}_{(1)}$ such that $\frac{\hat{y}_{(1)}+1-r}{3} -$

$[\frac{\hat{y}_{(1)}+1-r}{3}] = 0$, and $\hat{v}_1^* \neq \hat{v}_2^*$, we have

$$\varphi_3^*(\hat{v}_1^*, \hat{v}_2^*, v_3) = \begin{cases} v_3 & \text{if } v_3 \leq r + 3[\frac{\hat{y}_{(1)}-r}{3}] \\ \hat{y}_{(1)} & \text{if } r + 3[\frac{\hat{y}_{(1)}-r}{3}] < v_3 < r + 4[\frac{\hat{y}_{(1)}+1-r}{3}] - [\frac{\hat{y}_{(1)}-r}{3}] \\ \hat{y}_{(1)} + k & \text{if } v_3 \geq r + 4[\frac{\hat{y}_{(1)}+1-r}{3}] - [\frac{\hat{y}_{(1)}-r}{3}] \end{cases}$$

(4.4)

where k is defined by the following equation

$$k = \begin{cases} 1 & \text{if } [\frac{\hat{y}_{(1)}+1-r}{3}] < [\frac{\hat{y}_{(1)}+2-r}{3}] \\ 2 & \text{if } v_3 \geq \hat{y}_{(1)} + 2 \text{ and } [\frac{\hat{y}_{(1)}+1-r}{3}] = [\frac{\hat{y}_{(1)}+2-r}{3}] < [\frac{\hat{y}_{(1)}+3-r}{3}] \\ 3 & \text{if } v_3 \geq \hat{y}_{(1)} + 3 \text{ and } [\frac{\hat{y}_{(1)}+1-r}{3}] = [\frac{\hat{y}_{(1)}+3-r}{3}] \end{cases}$$

(4.5)

For $\hat{y}_{(1)}$ such that $\frac{\hat{y}_{(1)}+1-r}{3} - [\frac{\hat{y}_{(1)}+1-r}{3}] = 0$, and $\hat{v}_1^* = \hat{v}_2^*$, we have

$$\varphi_3^*(\hat{v}_1^*, \hat{v}_2^*, v_3) = \begin{cases} v_3 & \text{if } v_3 \leq r + 3[\frac{\hat{y}_{(1)}-r}{3}] \\ \hat{y}_{(1)} & \text{if } r + 3[\frac{\hat{y}_{(1)}-r}{3}] < v_3 < r + 3[\frac{\hat{y}_{(1)}+1-r}{3}] \\ \hat{y}_{(1)} + k & \text{if } v_3 \geq r + 3[\frac{\hat{y}_{(1)}+1-r}{3}] \end{cases}$$

(4.6)

With k defined according to Equation 4.5. Finally, for $\hat{y}_{(1)}$ such that $\frac{\hat{y}_{(1)}+1-r}{3} - [\frac{\hat{y}_{(1)}+1-r}{3}] > 0$, we have

$$\varphi_3^*(\hat{v}_1^*, \hat{v}_2^*, v_3) = \begin{cases} v_3 & \text{if } v_3 \leq r + 3[\frac{\hat{y}_{(1)}-r}{3}] \\ \hat{y}_{(1)} + 1 & \text{if } v_3 > r + 3[\frac{\hat{y}_{(1)}-r}{3}] \end{cases}$$

(4.7)

With k defined according to Equation 4.5. Let now be $\hat{y}_{(1)} < 58$. Equation 4.8 characterizes φ_3^* in this case.

$$\varphi_3^*(\hat{v}_1^*, \hat{v}_2^*, v_3) = \begin{cases} v_3 & \text{if } v_3 \leq 57 \\ 58 & \text{if } v_3 = 58 \\ 59 & \text{if } v_3 = 59 \\ 60 & \text{if } v_3 \geq 60 \end{cases}$$

(4.8)

PROOF: We start with Player 3's best announcement. In what follows, we show how we have gotten Expression 4.4. The other expressions can be derived in the same way. Note that $\hat{v}_3 > \hat{y}_{(1)} + k$ with k defined in Equation 4.5 is always a dominated strategy because for $\hat{v}_3 > \hat{y}_{(1)} + k$, we have

$$v_3 - r - 2 \cdot [\frac{\hat{y}_{(1)} + k - r}{3}] > v_3 - r - 2 \cdot [\frac{\hat{v}_3 - r}{3}]$$

(4.9)

Consider now Player 3's local payoff function

$$
\begin{aligned}
\pi_3 &= (v_3 - \hat{y}_{(1)} - k + [\frac{\hat{y}_{(1)} + k - r}{3}] + \\
&\quad 3 \cdot (\frac{\hat{y}_{(1)} + k - r}{3} - [\frac{\hat{y}_{(1)} + k - r}{3}])) \cdot 1_{\{\hat{v}_3 = \hat{y}_{(1)} + k\}} + \\
&\quad (\frac{1}{2}(v_3 - \hat{y}_{(1)}) + [\frac{\hat{y}_{(1)} - r}{3}] + 3 \cdot (\frac{\hat{y}_{(1)} - r}{3} - [\frac{\hat{y}_{(1)} - r}{3}])) + \\
&\quad \frac{1}{2}[\frac{\hat{y}_{(1)} - r}{3}]) \cdot 1_{\{\hat{v}_3 = \hat{y}_{(1)} > \hat{v}_{j \neq 3}\}} + \\
&\quad (\frac{1}{3}(v_3 - \hat{y}_{(1)}) + [\frac{\hat{y}_{(1)} - r}{3}] + 3 \cdot (\frac{\hat{y}_{(1)} - r}{3} - [\frac{\hat{y}_{(1)} - r}{3}])) + \\
&\quad \frac{2}{3}[\frac{\hat{y}_{(1)} - r}{3}]) \cdot 1_{\{\hat{v}_1^* = \hat{v}_2^* = \hat{v}_3\}} + \\
&\quad [\frac{\hat{y}_{(1)} - r}{3}] \cdot 1_{\{\hat{v}_3 < \hat{y}_{(1)}\}} \qquad\qquad (4.10)
\end{aligned}
$$

Consider $\hat{y}_{(1)}$ such that $\frac{\hat{y}_{(1)} + 1 - r}{3} - [\frac{\hat{y}_{(1)} + 1 - r}{3}] = 0$, and $\hat{v}_1 \neq \hat{v}_2$. If Player 3 announces $\hat{v}_3$ lower than $\hat{y}_{(1)}$ his payoff is equal to $[\frac{\hat{y}_{(1)} - r}{3}]$. If he announces $\hat{v}_3$ equal to $\hat{y}_{(1)}$ his payoff is equal to $(1/2) \cdot (v_3 - r - [(\hat{y}_{(1)} - r)/3])$. Finally, if he announces $\hat{v}_3$ equal to $\hat{y}_{(1)} + k$ his payoff is equal to $v_3 - r - 2 \cdot [(\hat{y}_{(1)} + k - r)/3]$. One can easily check that $[\frac{\hat{y}_{(1)} - r}{3}]$ is equal to or greater than $(1/2) \cdot (v_3 - r - [(\hat{y}_{(1)} - r)/3])$ and equal to or greater than $v_3 - r - 2 \cdot [(\hat{y}_{(1)} + k - r)/3]$ when $v_3 \leq r + 3[\frac{\hat{y}_{(1)} - r}{3}]$, and that $(1/2) \cdot (v_3 - r - [(\hat{y}_{(1)} - r)/3])$ is equal to or greater than $v_3 - r - 2 \cdot [(\hat{y}_{(1)} + k - r)/3]$ when $v_3 < r + 4[\frac{\hat{y}_{(1)} + 1 - r}{3}] - [\frac{\hat{y}_{(1)} - r}{3}]$.

It can be easily checked that this strategy is also the only one which satisfies the MUD property. Consider in fact the following example. Let be $v_3 = 80$ and $\hat{y}_{(1)} = 79$. According to Expression 4.7, Player 3's best announcement is $\hat{v}_3^* = 80$. Player 3 would get the same payoff by announcing $\hat{v}_3 = 81$. However, this announcement would violate the MUD property.

Worthy to be noted is the fact that, depending on $\hat{y}_{(1)}$, Player 3 will announce less or more than his value. Consider the following examples. Let be $v_3 = 90$ and $\hat{y}_{(1)} = 79$, then, according to Expressions 4.4 and 4.6, we have $\hat{v}_3^* = 81$. Now, let be $v_3 = 80$ and $\hat{y}_{(1)} = 80$. According to Expression 4.7, Player 3's best announcement is $\hat{v}_3^* = 81$. The reason for this is the absence of strict monotonicity of Player 3's payoff function in $\hat{y}_{(1)}$.

Now, we want to analyze the best announcement of the player who announces as the second one. He has already heard the announcement of

Player 1 and, since he is supposed to be rational, he anticipates the best announcement of Player 3. We distinguish between two cases: $v_2 \geq 58$ and $v_2 < 58$. We start with the case in which $v_2 < 58$. Remember that $(2/3) \cdot (\hat{v}_2 - 58) \leq \max\{0, v_2 - 58\}$, this means that the highest announcement he is allowed to do is 60 (since $[(60 - 58)/3] = 0$). In our analysis, we proceed as follows.

We show that when $\hat{v}_1^* \geq 60$, Player 2's best announcement is equal to $\hat{v}_2^* = v_2$. After that, we argue why $\hat{v}_2 = 58$ and $\hat{v}_2 = 59$ are non-optimal actions. Finally, we show in which cases Player 2's best announcement is equal to 60.

STEP 1: If $\hat{v}_1 \geq 60 \implies \hat{v}_2^* = v_2$.

Let $F_{\hat{v}_3^*}(\cdot)$ and $p_{\hat{v}_3^*}(\cdot)$ be the cumulative probability function and the probability function of Player 3's best announcement, respectively[6]. If $\hat{v}_1 > 60$, then, since the highest announcement Player 2 may make is 60, and since his payoff by announcing 60 is the same as by announcing his true value, Player 2's best announcement is $\hat{v}_2^* = v_2$. Now, we consider the case $\hat{v}_1 = 60$. In this case, Player 2's local payoff function is

$$
\begin{aligned}
\pi_2 = {} & (v_2 - r)1_{\{\hat{v}_2=60\}} \cdot F_{\hat{v}_3^*}(\hat{v}_2 - 1) + \\
& \frac{1}{2}(v_2 - r)1_{\{\hat{v}_2=60\}} \cdot p_{\hat{v}_3^*}(\hat{v}_2) + \\
& 1_{\{\hat{v}_2=60\}} \cdot (1 - F_{\hat{v}_3^*}(\hat{v}_2)) \quad\quad (4.11)
\end{aligned}
$$

Having heard $\hat{v}_1 = 60$, the highest announcement the last player will make is $\hat{v}_3 = 63$. That is, if $\hat{v}_1 = 60$ and $\hat{v}_2 = v_2$ then the expected payoff of the player who announces as the second one will be $(1 - F_{\hat{v}_3^*}(61))$, which depends on the fact whether he believes that v_1 is lower than r. Apart from his belief, announcing $\hat{v}_2 = 60$ there would result in a positive probability to get the item and, therefore, to realize a loss. This means that the expected payoff would be lower.

STEP 2: If $\hat{v}_1^* \leq 59 \implies \hat{v}_2 = 58$ and $\hat{v}_2 = 59$ are non-optimal actions.

If $v_3 > 58$ then Player 3 will announce at most $\hat{v}_3 = 60$, and that implies a negative expected payoff for Player 2, for $\hat{v}_2 = 58$ or $\hat{v}_2 = 59$. Therefore, he will never announce $\hat{v}_2 = 58$ or $\hat{v}_2 = 59$.

STEP 3: When Player 2 announces his own value and when he announces 60.

As we have seen, Equation 4.11 depends on whether Player 2 believes that v_1 is lower than r. That is, Player 2's best announcement depends on whether Player 2 believes that v_1 is lower, equal to or higher than

---

[6]Both the cumulative probability and the probability function can be calculated using Equations 4.2 until 4.7 and Equation 4.8

r. When $\hat{v}_1 \leq 59$, Player 2 can neither conclude that $v_1 \leq 57$ nor that $v_1 \geq 58$. In order to specify Player 2's beliefs, we need to know Player 1's equilibrium strategy. Given $\varphi_1^*(v_1)$, as specified in Figure 4.1, Player 2 is able, to build $\mu(\cdot \mid \hat{v}_1)$. That is, Player 2 is able to calculate $q = \sum_{v_1'=51}^{57} \mu(v_1' \mid \hat{v}_1)$ the probability that $v_1 \leq 57$ given $\hat{v}_1$ for each $\hat{v}_1$.

Using Equation 4.11 with different $F_{\hat{\mathbf{v}}_3^*}(\cdot \mid y_{(1)})$ (see Equation 4.2), we write down an optimal announcement depending on q for Player 2.

$$\varphi_2 = \begin{cases} v_2 & \text{if } \hat{v}_1 \geq 60 \text{ or } (v_2 - r)(817 - 567q) + 78(43 + 7q) \leq 0 \\ 60 & \text{otherwise} \end{cases}$$

Calculating q for each $\hat{v}_1^*$ and using the NSA condition for the probability-0 events, we get Equation 4.3. Note that this strategy is the only equilibrium strategy which satisfies the MUD property.

Consider now the case $v_2 \geq 58$. The local payoff function of Player 2 is equal to the following expression

$$\begin{aligned} \pi_2 = & ((v_2 - r - 2 \cdot [\frac{\hat{v}_2 - r}{3}]) \cdot F_{\hat{\mathbf{v}}_3^*}(\hat{v}_2 - 1) + (\frac{1}{2}(v_2 - r - [\frac{\hat{v}_2 - r}{3}]) + \\ & \frac{1}{2}[\frac{\hat{v}_2 - r}{3}]) \cdot p_{\hat{\mathbf{v}}_3^*}(\hat{v}_2) + \\ & [\frac{\hat{v}_2 + 1 - r}{3}] \cdot (1 - F_{\hat{\mathbf{v}}_3^*}(\hat{v}_2))) \cdot 1_{\{\hat{v}_2 > \hat{v}_1\}} + \\ & ((\frac{1}{2}(v_2 - r - [\frac{\hat{v}_1 - r}{3}]) \cdot F_{\hat{\mathbf{v}}_3^*}(\hat{v}_1 - 1) + \\ & (\frac{1}{3}(v_2 - r) \cdot p_{\hat{\mathbf{v}}_3^*}(\hat{v}_1) + \\ & [\frac{\hat{v}_1 + 1 - r}{3}] \cdot (1 - F_{\hat{\mathbf{v}}_3^*}(\hat{v}_1))) \cdot 1_{\{\hat{v}_2 = \hat{v}_1\}} + ([\frac{\hat{v}_1 - r}{3}] \cdot F_{\hat{\mathbf{v}}_3^*}(\hat{v}_1) + \\ & [\frac{\hat{v}_1 + 1 - r}{3}] \cdot (1 - F_{\hat{\mathbf{v}}_3^*}(\hat{v}_1))) \cdot 1_{\{\hat{v}_2 < \hat{v}_1\}} \end{aligned} \qquad (4.12)$$

Note that π_2 does not depend on v_1, for $v_2 \geq 58$. We have numerically maximized Expression 4.12. The exact solution is given in Appendix E. It is easy to check that Player 2's optimal announcement reported in Appendix E is the only one which satisfies the MUD property.

Now, we consider the best announcement for the player who announces as the first one. We start with the case $v_1 < 60$. Since in this case the player is not allowed to make any side payments, we have $\hat{v}_1 \leq 60$.

Let us introduce some notation. Let $p_{\hat{\mathbf{v}}_2^*, \hat{\mathbf{v}}_3^*}(\hat{\mathbf{v}}_2^* \leq x, \hat{\mathbf{v}}_3^* \leq y)$ be the conditional probability that Player 2's best announcement is equal to or lower than x, given that he has heard $\hat{v}_1$, and Player 3's best announcement is equal to or lower than y, given that he has heard $\hat{v}_1$ and $\hat{\mathbf{v}}_2^* \leq x$.

This probability function can be constructed using $p(v_i, v_j \mid v_k = y)$ as defined in Expression 4.1 and the best announcements of players who announce second and third, respectively. Let $p_{\hat{v}_2^*}(\hat{v}_2)$ be the conditional probability that Player 2's announcement is equal to $\hat{v}_2^* = \hat{v}_2$, given that he has heard $\hat{v}_1$. One can derive this probability from $\sum_{v_j} p(v_i, v_j \mid v_k = y)$ and from Player 2's announcement. With this notation, we can define Player 1's local payoff function.

$$
\begin{aligned}
\pi_1 \;=\; & (v_1 - r) \cdot 1_{\{\hat{v}_1 \geq r\}} \cdot p_{\hat{v}_2^*, \hat{v}_3^*}(\hat{\mathbf{v}}_2^* < \hat{v}_1, \hat{\mathbf{v}}_3^* < \hat{v}_1) + \\
& \frac{1}{2}(v_1 - r) \cdot 1_{\{\hat{v}_1 \geq r\}} \cdot (p_{\hat{v}_2^*, \hat{v}_3^*}(\hat{\mathbf{v}}_2^* = \hat{v}_1, \hat{\mathbf{v}}_3^* < \hat{v}_1) + \\
& p_{\hat{v}_2^*, \hat{v}_3^*}(\hat{\mathbf{v}}_2^* < \hat{v}_1, \hat{\mathbf{v}}_3^* = \hat{v}_1)) + \\
& \frac{1}{3}(v_1 - r) \cdot 1_{\{\hat{v}_1 \geq r\}} \cdot p_{\hat{v}_2^*, \hat{v}_3^*}(\hat{\mathbf{v}}_2^* = \hat{\mathbf{v}}_3^* = \hat{v}_1) + \\
& 1_{\{\hat{v}_1 = 60\}} \cdot p_{\hat{v}_2^*, \hat{v}_3^*}(\hat{\mathbf{v}}_2^* \leq \hat{v}_1, \hat{\mathbf{v}}_3^* = 61) + \\
& \Big(\sum_{\hat{v}_2 = 61}^{\overline{v}_2} \Big[\frac{\hat{v}_2 - r}{3}\Big] \frac{p_{\hat{v}_2^*}(\hat{v}_2)}{p_{\hat{v}_2^*}(\hat{\mathbf{v}}_2^* \geq 61)} \Big) \cdot p_{\hat{v}_2^*, \hat{v}_3^*}(\hat{\mathbf{v}}_2^* \geq 61, \hat{\mathbf{v}}_3^* \leq \hat{v}_1) + \\
& \Big(\sum_{\hat{v}_2 = 61}^{\overline{v}_2} \Big[\frac{\hat{v}_2 - r}{3}\Big] \frac{p_{\hat{v}_2^*, \hat{v}_3^*}(\hat{\mathbf{v}}_2^* = \hat{v}_2, \hat{\mathbf{v}}_3^* \leq \hat{v}_2)}{p_{\hat{v}_2^*}(\hat{\mathbf{v}}_2^* \geq 61)} + \\
& \sum_{\hat{v}_2 = 61}^{\overline{v}_2} \Big[\frac{\hat{v}_2 + 1 - r}{3}\Big] \frac{p_{\hat{v}_2^*, \hat{v}_3^*}(\hat{\mathbf{v}}_2^* = \hat{v}_2, \hat{\mathbf{v}}_3^* = \hat{v}_2 + 1)}{p_{\hat{v}_2^*}(\hat{\mathbf{v}}_2^* \geq 61)} \Big) \cdot \\
& p_{\hat{v}_2^*, \hat{v}_3^*}(\hat{\mathbf{v}}_2^* \geq 61, \hat{\mathbf{v}}_3^* \geq 61) + \\
& \Big(\sum_{\hat{v}_2 = 61}^{\overline{v}_2} \Big[\frac{\hat{v}_2 - r}{3}\Big] \frac{p_{\hat{v}_2^*}(\hat{v}_2)}{p_{\hat{v}_2^*}(\hat{\mathbf{v}}_2^* \geq 61)} \Big) \cdot p_{\hat{v}_2^*, \hat{v}_3^*}(\hat{\mathbf{v}}_2^* \geq 61, 60 \geq \hat{\mathbf{v}}_3^* > \hat{v}_1) + \\
& 1_{\{\hat{v}_1 < 60\}} \cdot p_{\hat{v}_2^*, \hat{v}_3^*}(\hat{v}_1 < \hat{\mathbf{v}}_2 = 60, \hat{\mathbf{v}}_3 = 61)
\end{aligned}
$$

Note that since $\hat{v}_1 \leq 60$, it must be $\overline{v}_2 = 84$ (see Appendix E). Indicating with C all terms in the expected payoff of Player 1 which do not depend on $\hat{v}_1$, we can rewrite π_1 as follows:

$$
\begin{aligned}
\pi_1 \;=\; & (v_1 - r) \cdot 1_{\{\hat{v}_1 \geq r\}} \cdot p_{\hat{v}_2^*, \hat{v}_3^*}(\hat{\mathbf{v}}_2^* < \hat{v}_1, \hat{\mathbf{v}}_3^* < \hat{v}_1) + \\
& \frac{1}{2}(v_1 - r) \cdot 1_{\{\hat{v}_1 \geq r\}} \cdot (p_{\hat{v}_2^*, \hat{v}_3^*}(\hat{\mathbf{v}}_2^* = \hat{v}_1, \hat{\mathbf{v}}_3^* < \hat{v}_1) + \\
& p_{\hat{v}_2^*, \hat{v}_3^*}(\hat{\mathbf{v}}_2^* < \hat{v}_1, \hat{\mathbf{v}}_3^* = \hat{v}_1)) + \\
& \frac{1}{3}(v_1 - r) \cdot 1_{\{\hat{v}_1 \geq r\}} \cdot p_{\hat{v}_2^*, \hat{v}_3^*}(\hat{\mathbf{v}}_2^* = \hat{\mathbf{v}}_3^* = \hat{v}_1) + \\
& 1_{\{\hat{v}_1 = 60\}} \cdot p_{\hat{v}_2^*, \hat{v}_3^*}(\hat{\mathbf{v}}_2^* \leq \hat{v}_1, \hat{\mathbf{v}}_3^* = 61) + \\
& 1_{\{\hat{v}_1 < 60\}} \cdot p_{\hat{v}_2^*, \hat{v}_3^*}(\hat{v}_1 < \hat{\mathbf{v}}_2^* = 60, \hat{\mathbf{v}}_3^* = 61) + C \qquad (4.13)
\end{aligned}
$$

Now, consider the case that $\hat{v}_1$ is equal to 60. Let $\pi_1(v_1 \mid \hat{v}_1)$ be Player 1's payoff when his value is equal to v_1 and his announcement is equal to $\hat{v}_1$. Using this notation and Equations 4.3 until 4.8, we get the following expression

$$
\begin{aligned}
\pi_1(v_1 \mid 60) = \ & (v_1 - r) \cdot (p_{v_2,v_3}(v_2 \leq 58, v_3 \leq 58) + \\
& \frac{1}{2} \cdot (p_{v_2,v_3}(v_2 = 59, v_3 \leq 58) + \\
& p_{v_2,v_3}(v_2 \leq 58, 59 \leq v_3 \leq 61)) + \\
& \frac{1}{3} \cdot p_{v_2,v_3}(v_2 = 59, 59 \leq v_3 \leq 60)) + \\
& p_{v_2,v_3}(v_2 \leq 58, v_3 \geq 62) + \\
& p_{v_2,v_3}(v_2 = 59, v_3 \geq 61) + C
\end{aligned} \tag{4.14}
$$

Using the same notation and the expressions mentioned above, we can write down Player 1's payoff when he announces his true value and $v_1 \leq 57$.

$$
\pi_1(v_1 \mid v_1) = p_{v_2,v_3}(v_2 \leq 59, v_3 \geq 62) + C \tag{4.15}
$$

It easy to check that $\pi_1(v_1 \mid v_1) > \pi_1(v_1 \mid 60)$ for every $v_1 \in \{51, \ldots, 57\}$. In the same way, for each $v_1 \in \{51, \ldots, 57\}$, one can check that $\pi_1(v_1 \mid v_1) \geq \pi_1(v_1 \mid \hat{v}_1)$ for each $\hat{v}_1 \in \{51, \ldots, 59\}$.

Now, we investigate the case in which $v_1 \in \{58, 59\}$. Using Expression 4.13 and observing that the all events are now conditioned on $v_{(1)} \geq 58$, it can be easily proven that $\pi_1(v_1 \mid 60) > \pi_1(v_1 \mid \hat{v}_1)$ for every $\hat{v}_1 \in \{51, \ldots, 59\}$.

We conclude considering the last case, namely $v_1 \geq 60$.

$$
\begin{aligned}
\pi_1 = \ & (v_1 - r - 2 \cdot [\frac{\hat{v}_1 - r}{3}]) \cdot p_{\hat{v}_2^*,\hat{v}_3^*}(\hat{v}_2^* < \hat{v}_1, \hat{v}_3^* < \hat{v}_1) + \\
& \frac{1}{2} \cdot (v_1 - r - [\frac{\hat{v}_1 - r}{3}]) \cdot (p_{\hat{v}_2^*,\hat{v}_3^*}(\hat{v}_2^* = \hat{v}_1, \hat{v}_3^* < \hat{v}_1) + \\
& p_{\hat{v}_2^*,\hat{v}_3^*}(\hat{v}_2^* < \hat{v}_1, \hat{v}_3^* = \hat{v}_1)) + \\
& \frac{1}{3} \cdot (v_1 - r - [\frac{\hat{v}_1 - r}{3}]) \cdot p_{\hat{v}_2^*,\hat{v}_3^*}(\hat{v}_2^* = \hat{v}_3^* = \hat{v}_1) + \\
& [\frac{\hat{v}_1 + 1 - r}{3}] \cdot p_{\hat{v}_2^*,\hat{v}_3^*}(\hat{v}_2^* \leq \hat{v}_1, \hat{v}_3^* = \hat{v}_1 + 1) + \\
& (\sum_{\hat{v}_2 = \hat{v}_1 + 1}^{\bar{v}_2} [\frac{\hat{v}_2 - r}{3}] \frac{p_{\hat{v}_2^*}(\hat{v}_2)}{p_{\hat{v}_2^*}(\hat{v}_2^* \geq \hat{v}_1 + 1)}) \cdot p_{\hat{v}_2^*,\hat{v}_3^*}(\hat{v}_2^* > \hat{v}_1, \hat{v}_3^* \leq \hat{v}_1) + \\
& (\sum_{\hat{v}_2 = \hat{v}_1 + 1}^{\bar{v}_2} [\frac{\hat{v}_2 - r}{3}] \frac{p_{\hat{v}_2^*,\hat{v}_3^*}(\hat{v}_2^* = \hat{v}_2, \hat{v}_3^* \leq \hat{v}_2)}{p_{\hat{v}_2^*}(\hat{v}_2^* \geq \hat{v}_1 + 1)} +
\end{aligned}
$$

$$\sum_{\hat{v}_2=\hat{v}_1+1}^{\bar{v}_2} [\frac{\hat{v}_2 + 1 - r}{3}] \frac{p_{\hat{\mathbf{v}}_2^*,\hat{\mathbf{v}}_3^*}(\hat{\mathbf{v}}_2^* = \hat{v}_2, \hat{\mathbf{v}}_3^* = \hat{v}_2 + 1)}{p_{\hat{\mathbf{v}}_2^*}(\hat{\mathbf{v}}_2^* \geq \hat{v}_1 + 1)}).$$

$$p_{\hat{\mathbf{v}}_2^*,\hat{\mathbf{v}}_3^*}(\hat{\mathbf{v}}_2^* > \hat{v}_1, \hat{\mathbf{v}}_3^* > \hat{v}_1) \qquad\qquad (4.16)$$

Note that $\bar{v}_2$ depends on $\hat{v}_1 + 1$. It is approximately equal to $\max\{84, \hat{v}_1 + 1\}$ if $\hat{v}_1 \leq 99$. The result of the numerical maximization of Expression 4.16 is shown by Figure 4.1. It is not difficult to check that this is the only equilibrium strategy of Player 1 which satisfies the MUD property. When all three players play according to the best announcements that we have defined above and Player 2 uses $\mu(\cdot \mid \hat{v}_1)$, we get the desired result. QED

Note that there are other perfect Bayesian equilibria which do not satisfy the MUD and NSA properties. Unfortunately, we have not been able to characterize all these equilibria. In what follows we give an example which shows that relaxing the MUD property not all equilibria yield the same outcome. Let $O(e, v_1, v_2, v_3)$ be the outcome function for some equilibrium e of the game defined above. That is, given that players are playing according to e, this function gives the end payoff of each player for each vector (v_1, v_2, v_3). If $e = e_{PBMN}$ is a perfect Bayesian equilibrium which satisfies the MUD and NSD properties and $(v_1, v_2, v_3) = (51, 86, 80)$, we know, from Proposition 1, that $\hat{v}_1^* = 51, \hat{v}_2^* = 78, \hat{v}_3^* = 79$, this means that $O(e_{PBMN}, 51, 86, 80)$ is equal to $(7, 7, 8)$, that is, Player 1, Player 2 and Player 3 obtain 7, 7 and 8, respectively. Now, we relax the MUD property and consider a perfect Bayesian equilibrium e_{PBN} in which Player 3 chooses always a lottery with probabilities $(1/2, 1/2)$ when he is indifferent between a lottery with probabilities $(1/2, 1/2)$ and a certain outcome. As one can easily check, being $v_3 = 80$ and having heard $\hat{y}_{(1)} = 78$, Player 3 is indifferent between announcing 78 and announcing 79. If he announces 79 he obtains 8 for sure, if he announces 78 he obtains 10 with probability 1/2 and 6 with probability 1/2. Without writing Player 3's strategy formally, we say that Player 3 will play as in Proposition 1 but, when he is indifferent between a lottery with probabilities $(1/2, 1/2)$ and a certain outcome, he will choose the lottery. Given Player 3's new strategy, Player 2 will maximize his local payoff function having heard $\hat{v}_1$. We do not want to completely characterize here Player 2's equilibrium strategy. We have calculated Player 2's best announcement for $\hat{v}_1 \leq 59$ (which is the highest value Player 1 may announce in our example). We have found that Player 2's best announcement is now equal to 75. But having heard $\hat{y}_{(1)} = 75$, Player 3's best announcement will be 76 (according to the new and the old strategy). That is, we have that $O(e_{PBN}, 51, 86, 80)$ is equal to $(6, 6, 10)$.

4.3   Bid-bargain Mechanism

In this section, we want to analyze the bid-bargain mechanism. As we have seen in Chapter 2, it is quite difficult to model the way in which experimental subjects actually played. They made bids and asks. Sometimes, they made an ask after a bid or, vice versa, they made a bid after an ask. In most of the cases, they made asks when they had relatively low values to find out whether the other players would bid something more. For this reason, and in order to write a manageable model, we disregard the asks.

The idealized model works as follows. Players draw their values from the distribution introduced in Section 4.1. We imagine that players use a thermometer. The thermometer has only 14 numbers (from 1 to 14) and at the beginning it displays 1. Bidders cannot reenter the bidding once they have exited.

At b, $(1 \leq b \leq 14)$, all active players have to announce simultaneously "Yes" or "No". If at least two players say "Yes" the temperature goes up to $b + 1$, otherwise if all active players say "No", the winner (who is chosen randomly with equal probabilities among the active players) has to pay 58 to the auctioneer and $b - 1$ to each loser, and the round ends. If only one player says " Yes ", he obtains the item, pays b to each loser, and the round ends. Note that bids higher than 14 are not serious since the highest value is 100. Saying "No" at 15, a player who has drawn 100 gets 14 for sure. On the contrary, saying "Yes", he would get only 12 (that is, 100-58-30). Since the side payments may not be higher than the realized payoff in each round, the following feasibility condition has to be satisfied: $2 \cdot b \leq v - r$ for each v.

Now, given that a player has a value v, which is the highest bid he will submit in equilibrium, given that at least another player is still active? Here, we examine symmetric equilibria in pure strategies. That is, we examine equilibria such that when $(n - 1)$ players play the same strategy, it is also optimal for the nth player to play the same strategy. Because of the feasibility condition $2 \cdot b \leq v - r$ for each v, we have that at $b = 0$ players who have values $v \leq 59$ have to announce "No". Let us assume that at the bid b $(b \geq 1)$ only two players are active. Player i has now to decide whether to say "Yes" or "No" at $b + 1$. Let be b_j^h, the highest bid until which Player j, having observed v_j, will say "Yes". Let us indicate with $\pi_{i,N}$ Player i's expected payoff when he says "No" at $b+1$, and with $\pi_{i,Y}$ Player i's expected payoff when he says "Yes" at $b+1$ and he will say "No" at $b+2$. Since b_j^h depends on v_j which is the realization of a random variable, we have to introduce the probability function of the random variable $\mathbf{b}_j^h$, highest bid until which Player j, having observed v_j, will say "Yes". Let $p_{\mathbf{b}_j^h}$ be the probability function of the random variable $\mathbf{b}_j^h$. To decide whether to say "Yes" or "No" at $b+1$, Player i has to compare the following two expressions

$$
\begin{aligned}
\pi_{i,N} \;=\;& (1/2)\cdot(v_i-b-58)\cdot p_{\mathbf{b}_j^h}(\mathbf{b}_j^h\le b)+(b+1)\cdot(1-p_{\mathbf{b}_j^h}(\mathbf{b}_j^h\le b)) \\
\pi_{i,Y} \;=\;& (v_i-2(b+1)-58)\cdot p_{\mathbf{b}_j^h}(\mathbf{b}_j^h\le b)+ \\
& (1/2)\cdot(v_i-(b+1)-58)\cdot p_{\mathbf{b}_j^h}(\mathbf{b}_j^h=b+1)+ \\
& (b+2)\cdot(1-p_{\mathbf{b}_j^h}(\mathbf{b}_j^h\le b+1))
\end{aligned}
$$

Note that this is true if $\pi_{i,Y}-\pi_{i,N}$ is monotonically decreasing in b, that is, only if it can never happen that $\pi_{i,Y}-\pi_{i,N}$ is negative at $b+1$ and positive at $b+i$, for some $i>1$. We proceed heuristically without checking at this stage whether $\pi_{i,Y}-\pi_{i,N}$ is monotonically decreasing. We will numerically check the monotonicity of $\pi_{i,Y}-\pi_{i,N}$ only for the suspected equilibrium.

If Player i knew the functional form of b_j^h, he could calculate the difference $\pi_{i,Y}-\pi_{i,N}$. As we have said above, the equilibria we are looking for are symmetric equilibria in pure strategies. In order to calculate these equilibria, we proceed in the following way. We assume that Player j's highest bid is $b_j^{h,1}$. Given $b_j^{h,1}$, Player i is able to calculate $\pi_{i,Y}-\pi_{i,N}$ for each b and for each v_i. That is, given $b_j^{h,1}$, Player i will be able to calculate his best highest bid, $b_i^{h,1}$, versus $b_j^{h,1}$. If $b_i^{h,1}$ coincides with $b_j^{h,1}$ then we stop, otherwise we assume that Player j's highest bid is $b_j^{h,2}=b_i^{h,1}$. Given $b_j^{h,2}$, Player i will be able to calculate $b_i^{h,2}$, Player i's best highest bid versus $b_j^{h,2}$. Again, if $b_i^{h,2}$ is equal to $b_j^{h,2}$ then we stop, otherwise we continue to iterate. If $b_i^{h,l}$ converges to $b_j^{h,l}$ for $l<\infty$ and if $\pi_{i,Y}-\pi_{i,N}$ is monotonically decreasing in b when players play according to $b_j^{h,l}$ then we have found an equilibrium. We have calculated an equilibrium assuming that $b_j^{h,1}$ is equal to $[\frac{v_j-58}{3}]$, the integer part of $\frac{v_j-58}{3}$ (see Figure 4.2). We have found that $b_i^{h,l}$ converges to $b_j^{h,l}$ for $l=4$. Moreover, we have numerically checked the monotonicity of $\pi_{i,Y}-\pi_{i,N}$ in b, when both players play according to $b^{h,l}$ with $l\in\{1,\ldots,4\}$. We have run the iterative process eight times, starting from $b_j^h=[\frac{v_j-r}{\overline{\alpha}_j}]$ with

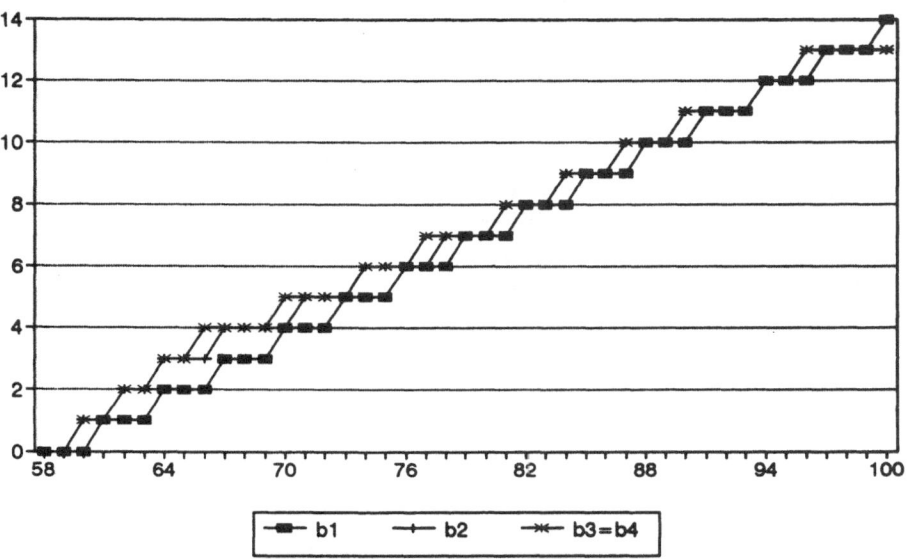

Figure 4.2: Equilibrium highest bid (when 2 players are active) starting from $b_j^h = \lceil \frac{v_j - 58}{3} \rceil$. Note that bl, with $l \in \{1, \ldots, 4\}$, is identically equal to $b_i^{h,l}$

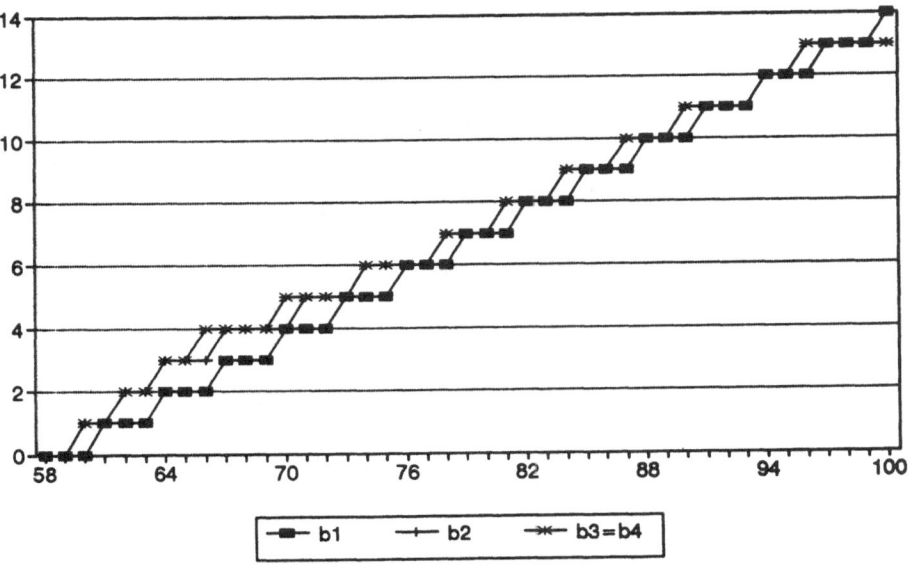

Figure 4.3: Equilibrium highest bid (when 2 players are active) starting from $b_j^h = \lceil \frac{v_j - r}{\overline{\alpha}_j} \rceil$, for $\overline{\alpha}_j > 3$. Note that bl, with $l \in \{1, \ldots, 4\}$, is identically equal to $b_i^{h,l}$

$\overline{\alpha}_j \in \{2, 2.4, 2.6, 4, 6, 8, 10, 15\}$. For all equilibria we have calculated, we have found that for $\overline{\alpha}_j \leq 3$, the equilibrium highest bid is equal to that one we have already discussed. However, for $\overline{\alpha}_j > 3$, the equilibrium highest bid is the one shown in Figure 4.3. Note that these equilibria are very similar. It has also to be noted that the convergence of the equilibrium highest bid is very rapid. We need at most six iterations to reach one of the two equilibria.

Now, we want to find the highest bid in equilibrium when all three players, i, j, k, are active. Extending the notation introduced above, we can define Player i's expected payoff when he says "No" at $b + 1$ and Player i's expected payoff when he says "Yes" at $b + 1$ and he will say "No" at $b + 2$.

$$
\begin{aligned}
\pi_{i,N} \; = \; & (1/3) \cdot (v_i - 58) \cdot p_{\mathbf{b}_j^h, \mathbf{b}_k^h}(\mathbf{b}_j^h \leq b, \mathbf{b}_k^h \leq b) + \\
& b \cdot (1 - p_{\mathbf{b}_j^h, \mathbf{b}_k^h}(\mathbf{b}_j^h \leq b, \mathbf{b}_k^h \leq b))
\end{aligned}
$$

$$
\begin{aligned}
\pi_{i,Y} \; = \; & (v_i - 2(b+1) - 58) \cdot p_{\mathbf{b}_j^h, \mathbf{b}_k^h}(\mathbf{b}_j^h \leq b, \mathbf{b}_k^h \leq b) + \\
& (1/3) \cdot (v_i - 58) \cdot p_{\mathbf{b}_j^h, \mathbf{b}_k^h}(\mathbf{b}_j^h = b+1, \mathbf{b}_k^h = b+1 + \\
& (1/2) \cdot (v_i - b - 58) \cdot (p_{\mathbf{b}_j^h, \mathbf{b}_k^h}(\mathbf{b}_j^h \leq b, \mathbf{b}_k^h = b+1) + \\
& p_{\mathbf{b}_j^h, \mathbf{b}_k^h}(\mathbf{b}_j^h = b+1, \mathbf{b}_k^h \leq b)) + (b+2) \cdot \\
& (1 - p_{\mathbf{b}_j^h, \mathbf{b}_k^h}(\mathbf{b}_j^h \leq b+1, \mathbf{b}_k^h \leq b+1)
\end{aligned}
$$

As above, we do not check at this point whether $\pi_{i,Y} - \pi_{i,N}$ is monotonically decreasing in b. Note that now we have to check that it can never happen that $\pi_{i,Y} - \pi_{i,N}$ is negative at $b+1$ and positive at $b+i$, for some $i > 1$, considering that at $b+i$ we can have either, again, all three players active or only two.

Following the same procedure used above, we set $b_j^{h,1} = b_k^{h,1} = [\frac{v-58}{3}]$, the integer part of $\frac{v-58}{3}$. Assuming now that Player j and Player k both play according to $b_j^{h,1}$ and $b_k^{h,1}$, respectively, we calculate Player i's best highest bid (see Figure 4.4). Iterating four times Player i's highest best bid, we find $b_i^{h,5} = b_j^{h,5} = b_k^{h,5}$. We have numerically checked that $\pi_{i,Y} - \pi_{i,N}$ is monotonically decreasing in b when all three players play according to $b^{h,l}$ with $l \in \{1, \ldots, 5\}$. That is, $b_i^{h,5}$ is an equilibrium highest bid, as defined above.

We have run the iterative process six times, starting from $b_j^{h,1} = b_k^{h,1} = [\frac{v-58}{\overline{\alpha}}]$ with $\overline{\alpha} \in \{2, 2.4, 2.8, 4, 8, 10\}$, and in all cases we have calculated we

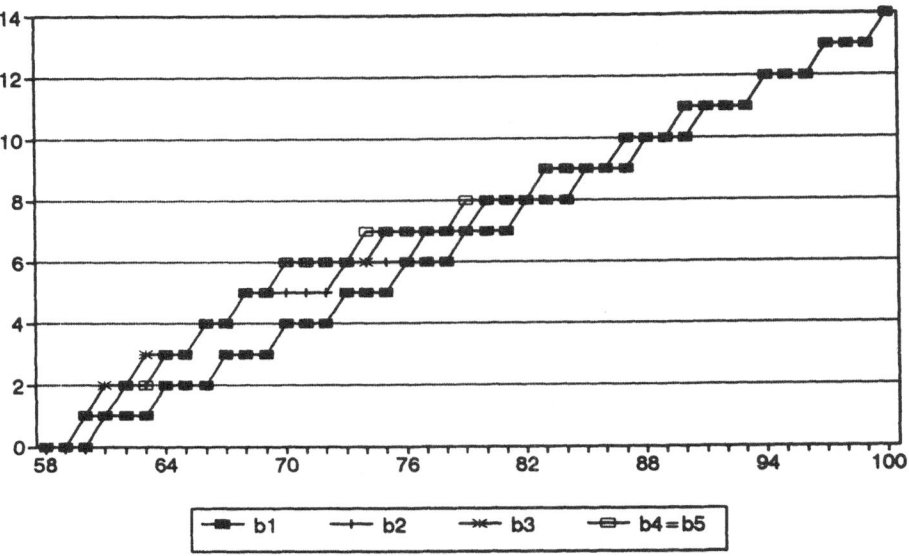

Figure 4.4: Equilibrium highest bid (when all 3 players are active). Note that bl, with $l \in \{1, \ldots, 5\}$, is identically equal to $b_i^{h,l}$

have found that the best highest bids converge to the equilibrium highest bid $b_i^{h,5}$ found above. We need at most seven iterations to converge to the equilibrium found above. Note that when only two players are active the highest bid is equal to or lower than in the case in which all three players are active. This is very intuitive. In fact, when all three players are active the probability that at least one of the other two players will say "Yes" is higher than in the case in which only two players are active. We have not checked whether there are asymmetric equilibria. Note also that the way in which we have approached the problem is not the most general. In fact, we have started to calculate our equilibrium without addressing the problem of equilibrium existence first. We have calculated the equilibria of the same model when there are 4 and 5 players and when the values are uniformly distributed, and we have found that such equilibria have the same structure as those we have seen above.

4.3.1 (Non-)Optimality

As we have already observed in Section 4.3, this mechanism is not completely optimal. There is in fact a positive probability that the player with the highest value does not obtain the item. This is due to the fact that the equilibrium highest bid function b_i^h is not strictly monotone in the players' values, that is, it may happen that players with different values have the same highest bid[7]. Here we want to calculate this probability for the equilibrium obtained starting from $b_j^{h,1} = b_k^{h,1} = [\frac{v-58}{3}]$ and show that the expected loss is very low.

The probability of non-optimality is equal to 2/3 of the probability that all three players have the same b_i^h and different values, plus 2/3 of the probability that all three players have the same b_i^h and two players have the same lowest values, plus 1/3 of the probability that all three players have the same b_i^h and two players have the same highest values, plus 1/2 of the probability that two players have the same highest b_i^h and different values. Formally, let n be the cardinality of the set of values with the same b_i^h and s the number of players with the same b_i^h and with different values, and let be $n \geq s$.

The probability that all three players have the same b_i^h ($i = 1, 2, 3$) and different values is equal to

$$\frac{1}{50^3 - 7^3} \cdot \binom{n}{s} \cdot s!$$

The probability that all three players have the same b_i^h ($i = 1, 2, 3$) and two players have the same values is equal to

$$\frac{1}{50^3 - 7^3} \cdot \binom{n}{s} \cdot 3$$

Finally, the probability that two players have the same highest bid b_i^h and different values is equal to

$$p_{\mathbf{v}_{\pi(1)}, \mathbf{v}_{\pi(2)}, \mathbf{v}_{\pi(3)}} (\mathbf{v}_{\pi(1)} = x, \mathbf{v}_{\pi(2)} = x + 1, \mathbf{v}_{\pi(3)} < x) \cdot \binom{n}{s} \cdot 3!$$

where $\pi(i)$ is a permutation of the index i, and x is such that the equilibrium highest bid for the players who drew x and $x + 1$, respectively, is the same and strictly higher than the equilibrium highest bid for the player who drew $x - 1$. With these definitions and using equilibrium highest bids shown in Figures 4.2 and 4.4, we can calculate the ex ante total probability that the player with the highest value does not get the item for the case that players play according to the equilibrium.

---

[7]Note that this does not result from the way in which we model players' behavior. On the contrary, this is due to the nature of the problem, namely to the fact that players are only allowed to transfer units.

Doing this calculation, we find out that the probability that the player with highest value does not obtain the item is just equal to 0.038. Moreover we can calculate the total expected loss[8], which is roughly equal to 0.114. Comparing this value with the optimal expected payoff we find that the highest expected loss is 0.38 per cent of the total expected payoff. As we have already pointed out, the suboptimality of this mechanism is due to the fact that players use integer bids. The suboptimality would completely disappear if players were allowed to use real bids.

4.4 Lattice Mechanism

In this section we want to analyze the lattice mechanism. Players draw their values from the distribution introduced in Section 4.1. As above, we indicate the integer part of x with $[x]$. In Subsection 2.4.2, we have already seen that subjects who use this mechanism proposed to play according to the partition shown in Table 4.1. That is, if a player draws a value v, say, equal to 74, he bids 59 and, if he obtains the object, he pays $[\frac{74-r}{3}]$ each of losing bidders. We have to observe that since values are and remain private information, it is individually rational to transfer $[(a_i - r_i)/3]$ instead of $[(v - r_i)/3]$, with $(r_i, a_i) \in \{(58, 58), (59, 71), (60, 81), (61, 91)\}$. Now, given that two players, say j and k, bid according to the proposed lattice, and given that they transfer always $[(a_i - r_i)/3]$ when they win having bid r_i, is it optimal for Player i to bid according to this lattice?

In what follows, we answer to this question. Let $p_{\mathbf{r}_j, \mathbf{r}_k}(\cdot, \cdot)$ be the joint probability function of $\mathbf{r}_j, \mathbf{r}_k$ (the bids submitted by Player j and Player k, respectively) given that Player i has observed v_i. If $v_i \leq 57$, then it is optimal for Player i bidding according to the lattice, it does

---

[8]We calculate this value through simulations.

Partition	Bids	Side Payments
$51, \ldots, 57$	< 58	$0, \ldots, 0$
$58, \ldots, 70$	58	$[\frac{58-58}{3}], \ldots, [\frac{70-58}{3}]$
$71, \ldots, 80$	59	$[\frac{71-59}{3}], \ldots, [\frac{80-59}{3}]$
$81, \ldots, 90$	60	$[\frac{81-60}{3}], \ldots, [\frac{90-60}{3}]$
$91, \ldots, 100$	61	$[\frac{91-61}{3}], \ldots, [\frac{100-61}{3}]$

Table 4.1: Lattice used by experimental subjects

not matter how the probability $p_{\mathbf{r}_j,\mathbf{r}_k}(\cdot,\cdot)$ looks like. In fact, if Player i bids more than 57 there would be a positive probability to realize a loss[9]. If Player i has drawn a value v_i equal to or higher than 58 and he knows that Player j and Player k bid according to the lattice, then he can calculate $p_{\mathbf{r}_j,\mathbf{r}_k}(\cdot,\cdot)$ using Expression 4.1, given in Section 4.1. Let be $s_i = [(a_i - r_i)/3]$ the side payment that Player i makes to each of losing bidders, when he bids r_i. To find Player i's optimal bid, we write his payoff function down when he has drawn $v_i \geq 58$ and bids r_i.

$$
\begin{aligned}
\pi_i(r_i, v_i) =\ & (v_i - 2s_i - r_i) \cdot p_{\mathbf{r}_j,\mathbf{r}_k}(\mathbf{r}_j \leq r_i - 1, \mathbf{r}_k \leq r_i - 1) + \\
& (\frac{1}{2} \cdot (v_i - 2s_i - r_i) + \frac{1}{2} \cdot s_i) \cdot \\
& (p_{\mathbf{r}_j,\mathbf{r}_k}(\mathbf{r}_j = r_i, \mathbf{r}_k \leq r_i - 1) + \\
& p_{\mathbf{r}_j,\mathbf{r}_k}(\mathbf{r}_j \leq r_i - 1, \mathbf{r}_k = r_i)) + \\
& (\frac{1}{3} \cdot (v_i - 2s_i - r_i) + \frac{2}{3} \cdot s_i) \cdot \\
& p_{\mathbf{r}_j,\mathbf{r}_k}(\mathbf{r}_j = r_i, \mathbf{r}_k = r_i) + \\
& \sum_{t=i+1}^{T-1} s_t \cdot (p_{\mathbf{r}_j,\mathbf{r}_k}(\mathbf{r}_j = r_t, \mathbf{r}_k = r_t) + \\
& p_{\mathbf{r}_j,\mathbf{r}_k}(\mathbf{r}_j = r_t, \mathbf{r}_k \leq r_{t-1}) + \\
& p_{\mathbf{r}_j,\mathbf{r}_k}(\mathbf{r}_j \leq r_{t-1}, \mathbf{r}_k = r_t)) + \\
& s_T(1 - p_{\mathbf{r}_j,\mathbf{r}_k}(\mathbf{r}_j \leq r_{T-1}, \mathbf{r}_k \leq r_{T-1})) \qquad (4.17)
\end{aligned}
$$

Numerical maximization of $\pi_i(r_i, v_i)$ gives us Player i's best reply, which is shown in Table 4.2. After four iterations of the best reply, we can find the equilibrium shown in Table 4.3.

Now, we want to show in a more general way how to construct the equilibria of the lattice mechanism. To do it, we distinguish between two

---

[9]Remember that the reserve price of our legitimate auction is equal to 58.

Partition	*Bids*	*Side Payments*
$51, \ldots, 57$	< 58	0
$58, \ldots, 73$	58	0
$74, \ldots, 86$	59	$\left[\frac{71-59}{3}\right]$
$87, \ldots, 100$	60	$\left[\frac{81-60}{3}\right]$

Table 4.2: Best reply when two players play according to the lattice used by experimental subjects

different cases. Firstly, given any partition of the set of the values and a system of bids, each for one element of the partition, we may want to find a system of side payments, each for one element of the partition, such that the resulting lattice is an equilibrium. That is, if a player has a value v_i which is contained in the ith partition element, then he has to make the corresponding bid and, if he is the winner, he has to make the due side payments. We will show below that one can find a solution to this problem. Secondly, given a system of side payments, we may want to find a partition such that the resulting lattice is an equilibrium. As we have already seen above, this cannot be done in general, at least when the set of values is given.

We start to consider the first case. Let $\{((\underline{v}_1, \overline{v}_1), r_1), \ldots, ((\underline{v}_l, \overline{v}_l), r_l)\}$ be a partition with $r_{i+1} - r_i = 1$, $\underline{v}_{i+1} < \overline{v}_{i+1}$ and $\underline{v}_{i+1} = \overline{v}_i + 1$ for each $i = 1, \ldots, l-1$. We want to find a system of side payments $\{s_1^*, \ldots, s_l^*\}$ such that if Player i has value $v_i \in \{\underline{v}_i, \ldots, \overline{v}_i\}$, then he will bid r_i and, if he wins the item, he will pay s_i^* to each losing bidder. In what follows, we assume that each subset of the partition has the same number of elements. This assumption together with the assumption of the uniform distribution of the values implies that $p_{\mathbf{r}_j, \mathbf{r}_k}(\mathbf{r}_j = r_i, \mathbf{r}_k = r_i) = p_{\mathbf{r}_j, \mathbf{r}_k}(\mathbf{r}_j = r_{i-1}, \mathbf{r}_k = r_{i-1})$. Moreover we fix the reserve value of the auctioneer equal to zero[10]. With these assumptions, we can write the difference of Player i's payoffs when he has drawn v_i and he bids r_i, and when he bids r_{i-1} as follows.

$$\pi(r_i, v_i) - \pi(r_{i-1}, v_i) = (v_i - r_i) \cdot (2 \cdot p_{\mathbf{r}_j, \mathbf{r}_k}(\mathbf{r}_j = r_{i-1}, \mathbf{r}_k = r_{i-1}) + p_{\mathbf{r}_j, \mathbf{r}_k}(\mathbf{r}_j = r_{i-1}, \mathbf{r}_k < r_{i-1}) + p_{\mathbf{r}_j, \mathbf{r}_k}(\mathbf{r}_j < r_{i-1}, \mathbf{r}_k = r_{i-1})) -$$

---

[10]These assumptions are not restrictive. The result can be easily generalized.

Partition	Bids	Side Payments
$51, \ldots, 57$	< 58	0
$58, \ldots, 72$	58	0
$73, \ldots, 86$	59	$\left[\frac{71-59}{3}\right]$
$87, \ldots, 100$	60	$\left[\frac{81-60}{3}\right]$

Table 4.3: Equilibrium Lattice (given the side payment function proposed by experimental subjects)

$$s_i \cdot (6 \cdot p_{\mathbf{r}_j, \mathbf{r}_k}(\mathbf{r}_j = r_{i-1}, \mathbf{r}_k = r_{i-1}) +$$
$$(7/2) \cdot p_{\mathbf{r}_j, \mathbf{r}_k}(\mathbf{r}_j < r_{i-1}, \mathbf{r}_k = r_{i-1}) +$$
$$(7/2) \cdot p_{\mathbf{r}_j, \mathbf{r}_k}(\mathbf{r}_j = r_{i-1}, \mathbf{r}_k < r_{i-1}) +$$
$$2 \cdot p_{\mathbf{r}_j, \mathbf{r}_k}(\mathbf{r}_j < r_{i-1}, \mathbf{r}_k < r_{i-1})) -$$
$$(1/3) \cdot p_{\mathbf{r}_j, \mathbf{r}_k}(\mathbf{r}_j = r_{i-1}, \mathbf{r}_k = r_{i-1}) -$$
$$p_{\mathbf{r}_j, \mathbf{r}_k}(\mathbf{r}_j < r_{i-1}, \mathbf{r}_k < r_{i-1}) -$$
$$(1/2) \cdot (p_{\mathbf{r}_j, \mathbf{r}_k}(\mathbf{r}_j < r_{i-1}, \mathbf{r}_k = r_{i-1}) +$$
$$p_{\mathbf{r}_j, \mathbf{r}_k}(\mathbf{r}_j = r_{i-1}, \mathbf{r}_k < r_{i-1})) +$$
$$s_{i-1} \cdot (2 \cdot p_{\mathbf{r}_j, \mathbf{r}_k}(\mathbf{r}_j < r_{i-1}, \mathbf{r}_k < r_{i-1}) +$$
$$p_{\mathbf{r}_j, \mathbf{r}_k}(\mathbf{r}_j < r_{i-1}, \mathbf{r}_k = r_{i-1}) +$$
$$p_{\mathbf{r}_j, \mathbf{r}_k}(\mathbf{r}_j = r_{i-1}, \mathbf{r}_k < r_{i-1}))$$

Note that $\pi(r_i, v_i) - \pi(r_{i-1}, v_i)$ is monotonically increasing in v_i. This implies that if $\pi(r_i, v_i) - \pi(r_{i-1}, v_i)$ is equal to or greater than zero for $v_i = \underline{v}_i$ then this is true for all $v_i > \underline{v}_i$.

Solving the inequality $\pi(r_i, v_i) - \pi(r_{i-1}, v_i) \geq 0$ in s_i, we have the first necessary condition

$$s_i^* \leq (1/(6 \cdot p_{\mathbf{r}_j, \mathbf{r}_k}(\mathbf{r}_j = r_{i-1}, \mathbf{r}_k = r_{i-1}) + (7/2) \cdot p_{\mathbf{r}_j, \mathbf{r}_k}(\mathbf{r}_j < r_{i-1}, \mathbf{r}_k = r_{i-1}) +$$
$$(7/2) \cdot p_{\mathbf{r}_j, \mathbf{r}_k}(\mathbf{r}_j = r_{i-1}, \mathbf{r}_k < r_{i-1}) + 2 \cdot p_{\mathbf{r}_j, \mathbf{r}_k}(\mathbf{r}_j < r_{i-1}, \mathbf{r}_k < r_{i-1}))) \cdot$$
$$((\underline{v}_i - r_i) \cdot (2 \cdot p_{\mathbf{r}_j, \mathbf{r}_k}(\mathbf{r}_j = r_{i-1}, \mathbf{r}_k = r_{i-1}) + p_{\mathbf{r}_j, \mathbf{r}_k}(\mathbf{r}_j = r_{i-1}, \mathbf{r}_k < r_{i-1}) +$$
$$p_{\mathbf{r}_j, \mathbf{r}_k}(\mathbf{r}_j < r_{i-1}, \mathbf{r}_k = r_{i-1}) - (1/3) \cdot p_{\mathbf{r}_j, \mathbf{r}_k}(\mathbf{r}_j = r_{i-1}, \mathbf{r}_k = r_{i-1})) -$$
$$p_{\mathbf{r}_j, \mathbf{r}_k}(\mathbf{r}_j < r_{i-1}, \mathbf{r}_k < r_{i-1}) - (1/2) \cdot (p_{\mathbf{r}_j, \mathbf{r}_k}(\mathbf{r}_j < r_{i-1}, \mathbf{r}_k = r_{i-1}) +$$
$$p_{\mathbf{r}_j, \mathbf{r}_k}(\mathbf{r}_j = r_{i-1}, \mathbf{r}_k < r_{i-1})) + s_{i-1} \cdot (2 \cdot p_{\mathbf{r}_j, \mathbf{r}_k}(\mathbf{r}_j < r_{i-1}, \mathbf{r}_k < r_{i-1}) +$$
$$p_{\mathbf{r}_j, \mathbf{r}_k}(\mathbf{r}_j < r_{i-1}, \mathbf{r}_k = r_{i-1}) + p_{\mathbf{r}_j, \mathbf{r}_k}(\mathbf{r}_j = r_{i-1}, \mathbf{r}_k < r_{i-1})))$$

for each $i = 2, \ldots, l$. Consider now $\pi(r_i, v_i) - \pi(r_{i+1}, v_i)$

$$\pi(r_i, v_i) - \pi(r_{i+1}, v_i) = -(v_i - r_i) \cdot (2 \cdot p_{\mathbf{r}_j, \mathbf{r}_k}(\mathbf{r}_j = r_i, \mathbf{r}_k = r_i) +$$
$$p_{\mathbf{r}_j, \mathbf{r}_k}(\mathbf{r}_j = r_i, \mathbf{r}_k < r_i) +$$
$$p_{\mathbf{r}_j, \mathbf{r}_k}(\mathbf{r}_j < r_i, \mathbf{r}_k = r_i)) -$$
$$s_{i+1} \cdot (6 \cdot p_{\mathbf{r}_j, \mathbf{r}_k}(\mathbf{r}_j = r_i, \mathbf{r}_k = r_i) +$$
$$(7/2) \cdot p_{\mathbf{r}_j, \mathbf{r}_k}(\mathbf{r}_j < r_i, \mathbf{r}_k = r_i) +$$
$$(7/2) \cdot p_{\mathbf{r}_j, \mathbf{r}_k}(\mathbf{r}_j = r_i, \mathbf{r}_k < r_i) +$$
$$2 \cdot p_{\mathbf{r}_j, \mathbf{r}_k}(\mathbf{r}_j < r_i, \mathbf{r}_k < r_{i-1})) +$$
$$(1/3) \cdot p_{\mathbf{r}_j, \mathbf{r}_k}(\mathbf{r}_j = r_i, \mathbf{r}_k = r_i) +$$
$$p_{\mathbf{r}_j, \mathbf{r}_k}(\mathbf{r}_j < r_i, \mathbf{r}_k < r_i) +$$

$$(1/2) \cdot (p_{\mathbf{r}_j, \mathbf{r}_k}(\mathbf{r}_j < r_i, \mathbf{r}_k = r_i) +$$
$$p_{\mathbf{r}_j, \mathbf{r}_k}(\mathbf{r}_j = r_i, \mathbf{r}_k < r_i)) -$$
$$s_i \cdot (2 \cdot p_{\mathbf{r}_j, \mathbf{r}_k}(\mathbf{r}_j < r_i, \mathbf{r}_k < r_i) +$$
$$p_{\mathbf{r}_j, \mathbf{r}_k}(\mathbf{r}_j < r_i, \mathbf{r}_k = r_i) +$$
$$p_{\mathbf{r}_j, \mathbf{r}_k}(\mathbf{r}_j = r_i, \mathbf{r}_k < r_i))$$

Note that $\pi(r_i, v_i) - \pi(r_{i+1}, v_i)$ is decreasing in v_i. That implies that if $\pi(r_i, v_i) - \pi(r_{i+1}, v_i)$ is equal to or greater than zero for $v_i = \bar{v}_i$ then this is true for all $v_i < \bar{v}_i$. Solving the inequality $\pi(r_i, v_i) - \pi(r_{i+1}, v_i) \geq 0$ in s_i, we have the second necessary condition

$$s_i \leq (1/(2 \cdot p_{\mathbf{r}_j, \mathbf{r}_k}(\mathbf{r}_j < r_i, \mathbf{r}_k < r_i) +$$
$$p_{\mathbf{r}_j, \mathbf{r}_k}(\mathbf{r}_j < r_i, \mathbf{r}_k = r_i) + p_{\mathbf{r}_j, \mathbf{r}_k}(\mathbf{r}_j = r_i, \mathbf{r}_k < r_i))) \cdot$$
$$(-(\bar{v}_i - r_i) \cdot (2 \cdot p_{\mathbf{r}_j, \mathbf{r}_k}(\mathbf{r}_j = r_i, \mathbf{r}_k = r_i) + p_{\mathbf{r}_j, \mathbf{r}_k}(\mathbf{r}_j = r_i, \mathbf{r}_k < r_i) +$$
$$p_{\mathbf{r}_j, \mathbf{r}_k}(\mathbf{r}_j < r_i, \mathbf{r}_k = r_i)) - s_{i+1} \cdot (6 \cdot p_{\mathbf{r}_j, \mathbf{r}_k}(\mathbf{r}_j = r_i, \mathbf{r}_k = r_i) +$$
$$(7/2) \cdot p_{\mathbf{r}_j, \mathbf{r}_k}(\mathbf{r}_j < r_i, \mathbf{r}_k = r_i) + (7/2) \cdot p_{\mathbf{r}_j, \mathbf{r}_k}(\mathbf{r}_j = r_i, \mathbf{r}_k < r_i) +$$
$$2 \cdot p_{\mathbf{r}_j, \mathbf{r}_k}(\mathbf{r}_j < r_i, \mathbf{r}_k < r_{i-1})) + (1/3) \cdot p_{\mathbf{r}_j, \mathbf{r}_k}(\mathbf{r}_j = r_i, \mathbf{r}_k = r_i) +$$
$$p_{\mathbf{r}_j, \mathbf{r}_k}(\mathbf{r}_j < r_i, \mathbf{r}_k < r_i) + (1/2) \cdot (p_{\mathbf{r}_j, \mathbf{r}_k}(\mathbf{r}_j < r_i, \mathbf{r}_k = r_i) +$$
$$p_{\mathbf{r}_j, \mathbf{r}_k}(\mathbf{r}_j = r_i, \mathbf{r}_k < r_i)))) \tag{4.18}$$

for each $i = 1, \ldots, l-1$. Writing down all these conditions for each $i = 1, \ldots, l$ we find $2l-1$ inequalities. There are $l-1$ of these inequalities that are redundant as can be easily observed, considering the expressions $\pi(r_i, v_i) - \pi(r_{i+1}, v_i)$ at $v_i = \bar{v}_i$ and $\pi(r_{i+1}, v_i) - \pi(r_i, v_i)$ at $v_i = \underline{v}_{i+1}$. These two expressions cannot be simultaneously greater than zero. It is therefore sufficient to solve the l remaining equations for each $\bar{v}_i$. This can be done recursively. One starts finding s_1^* which solves the equation $\pi(r_1, \underline{v}_1) = c$, where c is a constant. Then, given s_1^*, one finds s_2^*, which solves the equation $\pi(r_1, \bar{v}_1) - \pi(r_1, \underline{v}_2) = 0$. Then, given s_2^*, one can find with the same procedure s_3^*, and so continue until s_l^*. We have applied this procedure to calculate the side payments reported in Table 2.15.

Consider now the following system of bids and side payments $\{(r_1, s_1), \ldots, (r_l, s_l)\}$. We look for a partition with l subsets such that it is an equilibrium. That is, we want to find $\{(\underline{v}_1^*, \bar{v}_1^*), \ldots, (\underline{v}_l^*, \bar{v}_l^*)\}$, with $\underline{v}_i^* < \bar{v}_i^*$ and $\underline{v}_i^* = \bar{v}_{i-1}^* + 1$ for each i, such that, when Player i has a value $v_i \in \{\underline{v}_i^*, \bar{v}_i^*\}$, then he will bid r_i and, if he gets the item, he will pay s_i to each losing bidder. As above, we assume that each subset of the partition has the same number of elements and fix the reserve value of the auctioneer equal to zero.

To find this partition, we have to consider again $\pi(r_i) - \pi(r_{i+1})$ and $\pi(r_{i+1}) - \pi(r_i)$. Note that $p_{r_j, r_k}(\cdot, \cdot)$ depends now on the pair $(\underline{v}_i^*, \overline{v}_i^*)$ and has to be determined in equilibrium. As above, the $2l - 1$ inequalities can be reduced to a system of l equations which can be recursively solved. Table 4.3 shows the partition of equilibrium given the side payments proposed by experimental subjects. Note that if the set of values is given, as in our experiment, the partition which is an equilibrium for the assigned side payments, may not exist.

4.4.1   Optimality

As we have already observed in Subsection 2.5.7, the lattice mechanism is not optimal. There are two sources of non-optimality. The first one is that there is a positive probability that the player with the highest value does not obtain the item. The second one is the waste of resources caused by the use of bids higher than the reserve price. In this subsection, we want to show how to calculate the probability that the player with the highest value does not win the item and the expected loss caused by this fact. Summing the expected loss caused by bidding more than the reserve price, we get the total expected loss.

Let c_q be the cardinality of C_q, element of a partition of the set V of values, with $c_q \geq 2$ and $q = 1, \ldots, Q$. Let $\underline{v}_q$ and $\overline{v}_q$ be the lowest and highest element of C_q, respectively. Drawing three values from V, the probability that the player with the highest value does not win the item is equal to 2/3 of the probability that all three values are different and element of C_q, plus 2/3 of the probability that all three values are element of C_q and the lowest two are equal, plus 1/3 of the probability that all three values are element of C_q and the highest two are equal, plus 1/2 of the probability that the two highest values are different and element of C_q. Formally, the probability that the player with the highest value does not win the item is equal to

$$\sum_{q=1}^{Q} \left(\frac{2}{3} \cdot \frac{\binom{c_q}{3} \cdot 3!}{50^3 - 7^3} + \frac{2}{3} \cdot \frac{\binom{c_q}{2} \cdot 3}{50^3 - 7^3} + \frac{1}{3} \cdot \frac{\binom{c_q}{2} \cdot 3}{50^3 - 7^3} + \right.$$

$$\left. \frac{1}{2} \cdot \binom{c_q}{2} \cdot 3! \cdot p(\mathbf{v}_i = x, \mathbf{v}_j = y \neq x, \mathbf{v}_k \leq \overline{v}_{q-1}) \right)$$

with $x, y \in C_q$. Calculating this expression for the lattice reported in Table 4.3, we find out that it is equal to 0.14. Now, we want to calculate the expected loss caused by the fact that the player with the highest value does not win the item. We have to consider three events. The first one is the event that all three values are in C_q and all three are different. As we have seen above, the probability of non-optimality in this case is equal to $(2/3) \cdot \left(\begin{array}{c} c_q \\ 3 \end{array} \right) \cdot 3!)/(50^3 - 7^3)$. Since all three values are different, there is either an high loss or a low loss with equal probability. The expected high loss in each subset C_q is given by the following expression

$$\frac{1}{2} \cdot \frac{2}{3} \cdot \sum_{j=1}^{c_q-2} \frac{(c_q - j - 1) \cdot j \cdot 3!}{50^3 - 7^3} \cdot (c_q - j)$$

Note that $\sum_{j=1}^{c_q-2}(c_q - j - 1) \cdot j$ is exactly equal to $\left(\begin{array}{c} c_q \\ 3 \end{array} \right)$. The expected low loss in each subset C_q is given by the following expression

$$\frac{1}{2} \cdot \frac{2}{3} \cdot \sum_{j=1}^{c_q-2} \frac{(\sum_{i=1}^{j} i) \cdot 3!}{50^3 - 7^3} \cdot (c_q - j - 1)$$

Note again that $\sum_{j=1}^{c_q-2}(\sum_{i=1}^{j} i)$ is exactly equal to $\left(\begin{array}{c} c_q \\ 3 \end{array} \right)$. The second event we have to consider is that all three values are in the same C_q and that two of these are equal. In this case we can have either a loss with probability $2/3$ or a loss with probability $1/3$. As we have seen above, the probability of non-optimality in this case is equal to $(2/3 + 1/3) \cdot (\left(\begin{array}{c} c_q \\ 2 \end{array} \right) \cdot 3)/(50^3 - 7^3)$. In this case, the expected loss in each subset C_q is given by

$$(\frac{1}{3} + \frac{2}{3}) \cdot \sum_{j=1}^{c_q-1} \frac{j \cdot 3}{50^3 - 7^3} \cdot (c_q - j)$$

with $\sum_{j=1}^{c_q-1} j$ equal to $\left(\begin{array}{c} c_q \\ 2 \end{array} \right)$.

Finally, we have to consider the event that the highest two values are drawn from the same C_q and the other one is drawn from C_p, with $p < q$. As we have seen above, the probability that the player with the highest value does not win the item in this case is given by

$$\frac{1}{2} \cdot \left(\begin{array}{c} c_q \\ 2 \end{array} \right) \cdot 3! \cdot p(v_i = x, v_j = y \neq x, v_k \leq \bar{v}_{q-1})$$

with $x, y \in C_q$. The expected loss caused by this event is given by

$$\frac{3!}{2} \cdot \sum_{h=1}^{c_q-1} \cdot h \cdot p(v_i = x, v_j = y \neq x, v_k \leq \bar{v}_{q-1}) \cdot (c_q - h)$$

Summing the expected losses for each event and for each subset of a partition, we get the total expected loss caused by a non-optimal allocation of the item. We have calculated this value for the equilibrium lattice generated by the side payment function proposed by experimental subjects (see Table 4.3), and we have found out that it is equal to 0.98382.

Now, we want to calculate the expected loss caused by bidding more than the reserve price. Let be $r_q = r + q - 1$, with $q = 1, \ldots, Q$. Then we have that the expected loss is equal to

$$\sum_{q=1}^{Q} (q-1) \cdot p_{\mathbf{v}_{(1)}}(\underline{v}_q \leq v_{(1)} \leq \overline{v}_q)$$

where $v_{(1)} = \max\{v_1, v_2, v_3\}$ and $p_{\mathbf{v}_{(1)}}$ is the probability that $v_{(1)}$ is in the set $\{\underline{v}_q, \ldots, \overline{v}_q\}$. Calculating the expected loss for the equilibrium lattice shown in Table 4.3, we find that it is roughly equal to 1.55. Summing up, we have found that the expected loss of the lattice mechanism is roughly equal to 2.53.

4.4.2   The Best Lattice With Four Intervals

As we have seen above, there are two sources of non-optimality for the lattice mechanism: one is due to the fact that there is a positive probability that the player with highest value does not obtain the item. The other one is due to the bidding more than the reserve price. Unfortunately, it is not possible to minimize these two effects simultaneously. In fact, if one wants to minimize the probability that the player with the highest value does not win, one has to use partitions with many subsets, all with the same number of elements[11] but this implies a waste of money by bidding. On the other side, if one wants to avoid waste of money one has to use partitions with few subsets or, at least, partitions which have few elements where the probability is more concentrated and the losses are low. Taking into consideration these two opposite effects, we have

---

[11]Remember that $\binom{n_1 + n_2}{s} > \binom{n_1}{s} + \binom{n_2}{s}$, that is, an higher number of intervals within a partition reduces the probability that the player with the highest value does not get the item. Moreover, given two partitions with the same number of intervals, say two, with cardinalities n_1, n_2 for the first partition and n_3, n_4 for the second ($n_1 + n_2 = n_3 + n_4$) and $n_1 = n_2$, we have $\binom{n_1}{s} + \binom{n_2}{s} < \binom{n_3}{s} + \binom{n_4}{s}$. That is, given two partitions with the same number of intervals, the lowest probability of non-optimality is associated to the partition which has the same number of elements in each interval.

numerically calculated the lattice with 4 subsets which has the lowest total expected loss. The loss associated with the best lattice is roughly equal to 2.34. As we have already seen above, given any partition, we can calculate the side payments function, such that this lattice is an equilibrium. Adapting Expression 4.18 to take into account asymmetry of the classes, the reserve price greater than zero, and setting the side payment of the first two intervals equal to zero, we have found the side payments of equilibrium which are shown in Table 4.4.

Note that in Table 4.4 the number of elements of the intervals is not the same. The reason is that different elements have different "costs". More symmetry in the partition elements would reduce the loss due the fact that the player with the highest value does not get the item, but it would increase the loss caused by bidding more than reserve price.

4.5 First-Price Auction Mechanism

As we have already seen, experimental subjects used the first-price auction mechanism in the last three rounds of Auction 1. This mechanism works as follows. The player who submits the highest bid has to pay his bid minus the reserve price r divided by three to each loser and r to the auctioneer, if his bid is higher than r. As we have already observed, this mechanism is not the same as the mechanism proposed by McAfee and McMillan (1992), although it is very similar. In their mechanism, the player who submits the highest bid has to pay his bid minus the reserve price r divided by two to each loser and r to the auctioneer, if his bid is higher than r. We are looking for a symmetric Nash equilibrium of the mechanism used by experimental subjects. We will find it calculating a symmetric Nash equilibrium of a class of first-price auctions which contains the mechanism used by experimental subjects as well as the mechanism proposed by McAfee and McMillan (1992). We will

Partition	Bids	Side Payments
$51, \ldots, 57$	< 58	0
$58, \ldots, 77$	58	0
$78, \ldots, 91$	59	5.7
$92, \ldots, 100$	60	8.4

Table 4.4: Best lattice with four intervals for the set $\{51, \ldots, 100\}$

study the problem assuming, without loss of generality, that the values are independently drawn from the same distribution function F, with $F : [\underline{v}, \overline{v}] \to [0, 1]^{12}$.

Consider the following class of mechanisms. The player who submits the highest bid has to pay r to the auctioneer and his bid minus the reserve price r divided by $(n + k)$, where k is any real number greater than $-n$, to each loser, if his bid is higher than r. Note that for $k = -1$, we have the mechanism proposed by McAfee and McMillan (1992), and that for $k = 0$, we have the mechanism used by experimental subjects. Note also that for $k = -n$ and for $k = \infty$, it is an equilibrium to bid the reserve price and infinity (or the highest allowed bid), respectively, if $v \geq r$, and to bid v if $v < r$. By the revelation principle, the restriction to direct revelation mechanisms is without loss of generality: any Nash equilibrium outcome of any game which determines the sole bidder and side payments will also be a Nash equilibrium outcome of some direct revelation game in which the bidders report their valuations truthfully.

Proposition 2 *Given any $k > -n$, the following mechanism is incentive-compatible and efficient. Before the auction, the cartel members report their valuations to the mechanism. If no report exceeds r, the cartel does not bid in the auction. If at least one bid exceeds r, the bidder making the highest report v obtains the item and pays a total of*

$$\frac{(n-1)}{(n+k)} \cdot (P_k(v) - r) + r \tag{4.19}$$

where

$$P_k(v) = F(v)^{-n} \times \int_r^v (u - r)(n + k)F(u)^{n-1}f(u)du + r \tag{4.20}$$

Each losing bidder receives from the winner $(P_k(v) - r)/(n + k)$, and the seller receives r.

PROOF: Since $P_k(v)$ is monotonically increasing for each $k > -n$, the mechanism is clearly efficient. To check the incentive compatibility, consider w the value reported by a bidder with value v. Since a report of $w < r$ will never win and results in a constant profit, we have to consider only the case $w \geq r$. The payoff function of a player is given by

$$\pi = (v - \frac{n-1}{n+k}(P_k(w) - r) - r) \cdot F(w)^{n-1} +$$

---

[12]This result can be easily generalized to our case.

$$[1 - F(w)^{n-1}] \int_w^{\bar{v}} \frac{P_k(u) - r}{n + k} \cdot \frac{(n-1) \cdot F(u)^{n-2} f(u)}{1 - F(w)^{n-1}} du =$$

$$(v - \frac{n-1}{n+k}(P_k(w) - r) - r) \cdot F(w)^{n-1} +$$

$$\int_w^{\bar{v}} \frac{P_k(u) - r}{n + k} \cdot (n-1) \cdot F(u)^{n-2} f(u) du$$

First order conditions imply

$$\frac{\partial \pi}{\partial w} = (-P_k'(w)) \cdot F(w)^{n-1} +$$
$$(v - r) \cdot (n + k) \cdot F(u)^{n-2} f(u) -$$
$$n \cdot (P_k(w) - r) \cdot F(u)^{n-2} f(u) \qquad (4.21)$$

Since $\partial^2 \pi / \partial w \partial v \geq 0$, and since this condition assures the pseudoconcavity of π (see Matthews (1990)), this implies that the incentive compatibility is characterized by the following condition

$$\left. \frac{\partial \pi}{\partial w} \right|_{w = v} = 0$$

Replacing Expression 4.20 in 4.21 and setting $w = v$, one gets the desired result.   QED

Note that when k increases, $P_k(v)$ increases linearly but the fraction of $(P_k(v) - r)$ paid to each losing bidder decreases (linearly) in the same proportion, that is, the total amount paid by the player who announces the highest value is always the same for each k. This result is not surprising. McAfee and McMillan (1992) have shown that an efficient cartel mechanism has the property that the winner transfers to each of the losers an amount equal to $E[v_{(2)} - r \mid v_{(1)} \geq r]/n$, where $v_{(j)}$ represents the jth order statistic and the expectation is taken over the distribution of the highest valuation. The winner's expected rent is this amount plus the rent he would have earned if the auction had been non-cooperative. Note also that for $k = -1$, we obtain McAfee and McMillan result (see Theorem 1).

These mechanisms are not unique. The cartel can set up auctions like the one proposed above of its own as the following corollary shows.

Corollary 2 *Given any $k > -n$, the following mechanism is also Bayes-Nash implementable and efficient: the players organize prior sealed bid first-price auction among themselves. If the highest bid in these prior auction exceeds r, the winner then bids r in the legitimate auction and*

pays each of losers one (n+k)*th of the difference between his bid in the prior auction and r.*

To prove this corollary, note that, in each of the new mechanisms, bidding $P_k(v)$ is an equilibrium because, if all others bid $P_k(v)$, bidding $P_k(v)$ in the new mechanism has the same effect as responding honestly in the direct mechanism. Since $P_k(v)$ is monotonically increasing, these mechanisms are efficient[13].

Note that the theoretical prediction we have provided in Subsection 2.5.8 is an approximation of the true theoretical prediction. In fact, we have just rounded off Expression 4.20 disregarding the fact that experimental subjects may transfer only units. Because of the meager number of experimental observations, we consider only this approximation.

As we have already observed in Chapter 2, the mechanism used by experimental subjects is not completely optimal when we allow only integer bids. That is, there is a positive (but small) probability that the player with the highest value does not get the item. Here, we do not explicitly calculate the ex ante expected loss due to this fact. This can be done using the same formula given in Subsection 4.3.1. We have calculated the ex ante expected loss per player and have found out that it is just equal to 0.002.

4.6   Summing Up

In this chapter, we have analyzed the mechanisms used by experimental subjects from a strategic point of view. Table 4.5 sums up the ex ante expected payoff of each mechanism and his ex ante expected loss in percentage. Remember that, for each player, the ex ante expected payoff of an optimal mechanism is 10.04.

The announcement mechanism has many equilibria. Only one of these satisfies the MUD and the NSD properties. We have calculated the ex ante expected payoff of this equilibrium. This mechanism is not optimal. The suboptimality persists even if we allow continuous announcements.

The bid-bargain mechanism, on the contrary, is almost optimal. We have calculated the expected loss that the players suffer when they play

---

[13]The result given in Proposition 2 holds also for $k < -n$. In this case, the player who announces the highest value will pay r to the auctioneer and $(P_k(v)-r)/(n+k)$ to each loser. Note that numerator as well as the denominator of the last expression are negative. This means that considering the class of first-price auctions with $k < -n$, the winner has to be the player who submits the lowest bid.

Mechanism	Expected payoff	Expected loss
Announcement Mechanism	8.84	11.9 %
Bid-bargain Mechanism	10.002	0.38 %
Lattice Mechanism	9.16	8.7 %
First-price auction Mechanism	10.038	0.02 %

Table 4.5: Ex ante expected payoffs and losses

such a mechanism and we have seen that it is relatively low. This mechanism becomes completely optimal by allowing continuous bids.

We have shown how to construct an equilibrium for the lattice mechanism both when the partition is given and when the side payment function is given. We have calculated the total expected loss which players suffer when they play this mechanism. We have calculated the best lattice for the set $\{51, \ldots, 100\}$ (restricting ourselves to the partitions with four intervals), that is the lattice which minimizes the loss suffered by players.

Finally, we have analyzed the first-price auction mechanism in the version proposed by experimental subjects. Because of the units problem, this model is not completely optimal. However, the ex ante expected loss is lower than the loss suffered with the bid-bargain mechanism.

Chapter 5

Two Extensions

In this chapter we consider two interesting extensions. The case of the lattice mechanism with continuous bids is investigated in Section 5.1. In Section 5.2, we consider the case of a coalition of two players which bids non-cooperatively versus an individual bidder.

5.1 Lattice With Continuous Bids

As we have already seen in Subsection 2.5.7, the lattice mechanism is non-optimal since the players have to bid integer numbers. In this section, we assume that the values are drawn from an interval $[\underline{v}, \overline{v}]$ and the bids may be real numbers. By allowing to submit real bids, we can minimize the loss caused by bidding more than the reserve price and at the same time completely eliminate the other source of inefficiency, namely that the bidder with the highest value may not obtain the item. We find one bid and one side payment for each value. In this way, the player with the highest value obtains the item[1].

We prove that for each monotonically strictly increasing bid function $B : [\underline{v}, \overline{v}] \to [0, r + \varepsilon]$, with ε small enough, there is a side payment function G and a bidding function $B(v)$ such that it is a Nash equilibrium bidding $B(v)$ in the main auction and, if one gets the item, paying G to each losing bidder. In what follows we assume that all values are independently drawn from the same distribution $F : [\underline{v}, \overline{v}] \to [0, 1]$, with differentiable density function $f(\cdot)$[2]. Let be $B : [\underline{v}, \overline{v}] \to [0, r + \varepsilon]$ any

---

[1]Note that all we need to reduce the inefficiency of the lattice mechanism is to allow players to submit real bids. We assume that the types are continuous only because it makes the problem more tractable.

[2]Although we prove our result under the assumption of independence, the result is true when we assume that each single variable is uniform distributed and the

monotonically strictly increasing and differentiable function such that
$B(v) = v$ for $v < r$ and $B(r) = r$, and let G be differentiable. As above,
we require that side payments may not be higher than the realized payoff
in each round. We can prove the following proposition.

Proposition 3 *One Nash equilibrium is to bid $B(v)$ in the legitimate
auction. If $B(v) \geq r$ for at least one player, then the player who made
the highest bid $B(v) = b$ will pay $\frac{G(B^{-1}(b))}{n-1}$ to each loser and $B(v)$ to the
auctioneer, where*

$$G(v) = F(v)^{-n} \cdot \int_r^v ((u-B(u))(n-1)F(u)^{n-1}f(u) - B'(u)F(u)^n)du \quad (5.1)$$

PROOF: If $v < r$, any bid b equal to or higher than r would violate the
feasibility condition $(v - b) \geq G(B^{-1}(b))$. Let be $v \geq r$. Since $B(\cdot)$ is
assumed to be strictly increasing, it is invertible. That is, choosing which
bid b' to submit is equivalent to choose which value to report. Let w be
equal to $B^{-1}(b')$. Assuming that the other $n - 1$ bidders are bidding
according to $B(\cdot)$, we can write the payoff function of a player

$$\begin{aligned}
\pi &= (v - B(w) - G(w)) \cdot F(w)^{n-1} + \\
&\quad [1 - F(w)^{n-1}] \int_w^{\bar{v}} \frac{G(u)}{n-1} \cdot \frac{(n-1) \cdot F(u)^{n-2}f(u)}{1 - F(w)^{n-1}} du = \\
&\quad (v - B(w) - G(w)) \cdot F(w)^{n-1} + \int_w^{\bar{v}} \frac{G(u)}{n-1} \cdot (n-1) \cdot F(u)^{n-2}f(u)du
\end{aligned}$$

First order conditions imply

$$\begin{aligned}
\frac{\partial \pi}{\partial w} &= (-B'(w) - G'(w)) \cdot F(w)^{n-1} + \\
&\quad (v - B(w) - G(w)) \cdot (n-1) \cdot F(u)^{n-2}f(u) - \\
&\quad G(w) \cdot F(u)^{n-2}f(u) \quad (5.2)
\end{aligned}$$

Since $\partial^2 \pi / \partial w \partial v \geq 0$, and since this condition assures the pseudoconcavity of π (see Matthews (1990)), this implies that the incentive compatibility is characterized by the following condition

$$\left. \frac{\partial \pi}{\partial w} \right|_{w=v} = 0$$

---

maximum of the n random variables has to be equal to or higher than r. In fact, if a
player observes a value which is lower than r then it is optimal for him to bid zero,
otherwise he is concerned only with the distribution of the maximum of the $n - 1$
variables and this does not depend on his observation.

Using Expression 5.1 in 5.2 and setting $w = v$, one gets the desired result. QED

Note that in the proof we use heavily the fact that B has an inverse. This mechanism is supposed to be useful when players may not communicate after that they have learned their values. If they could communicate, there would be no reason to suppose that they play a non-optimal mechanism. One could also think that although, as in our experiment, players are allowed communicate, the communication costs are very high.

Another justification to use this mechanism is that there is a positive probability that the auctioneer does not sell the object when he observes that the highest bid is equal to the reserve price. Estimating the price at which the auctioneer is really ready to sell the item one can fix ε appropriately.

Finally, note that when B is constant equal to the reserve price r, we have the optimal mechanism proposed by McAfee and McMillan (1992) as a special case.

5.2 Two-player Coalition

As we have seen in Subsection 2.4.4, there is an auction in which two players collude versus another one in the last two rounds. The theoretical solution of the announcement mechanism is not trivial in this case. In fact, one has to determine simultaneously the optimal announcement for each player and the optimal bid. We do not want to address this problem here. Since experimental subjects have always announced their true values in this auction, we assume that the values of the members of the coalition are common knowledge within the coalition. We also assume that the single player knows this. In this way, we can restrict our attention to the bids submitted by the coalition and the single player. Analyzing this case, we follow Marshall et al. (1994)[3]. To simplify our analysis, we assume, as in Marshall et al., that the values are drawn from an interval. As above, we maintain the assumption that the values are jointly uniformly distributed and that the probability that the maximum of the three values is lower than 58 is equal to zero[4]. Let $f(v_1, v_2, v_3)$ be the joint density function of three jointly uniformly distributed variables

---

[3]They consider both the case in which a coalition play versus another coalition, and the case in which a coalition play versus individual non collusive bidders. With three players, as in our set-up, the two cases coincide.

[4]Marshall et al. make a more standard assumption that the values of each player are independently drawn by uniform distributions on the interval [0, 1].

which have a maximum equal to or greater than 58 and equal to or lower than 100. We have

$$f(v_1, v_2, v_3) = \begin{cases} \frac{1}{50^3 - 8^3} & \text{if } v_{(1)} \geq 58 \\ 0 & \text{otherwise} \end{cases} \tag{5.3}$$

where $v_{(1)} = \max\{v_1, v_2, v_3\}$.

Given this density, we can, using the Bayes rule, derive the ex post distribution function of the maximum of two random variables given that we have observed one variable, and the distribution function of one random variable, given that we have already observed two variables[5].

When the highest value of the coalition, respectively the value drawn by the single player are lower than 58 it is optimal not to submit any bid. Therefore, we consider only the case in which v is equal to or greater than 58. Le Brun (1991) has shown that a Nash equilibrium exists in this case and it is unique. Moreover he has shown that the bid functions are strictly monotone. We subscribe with S and C the single player and the coalition, respectively. Since $b_i(\cdot)$ is monotone, it is invertible. Let φ_i be the inverse of the bid function b_i. It is an obvious necessary condition for the pair (φ_C, φ_S) to be a Nash equilibrium that they have a common support $[58, b^*]$, where b^* is the bid submitted when $v = 100$. Let be $b_C = b_C(v)$ the Nash equilibrium bid submitted by the coalition when its highest value is v. Hence b_C is generated by

$$b_C = \arg\max_b \{(v - b) \cdot \frac{\varphi_S(b) - 58}{50}\} \tag{5.4}$$

The first-order condition generates the following differential equation

$$(\varphi_C(b) - b) \cdot \varphi_S'(b) = \varphi_S(b) - 58 \tag{5.5}$$

The corresponding first-order condition for the single player generates the following differential equation

$$2 \cdot (\varphi_S(b) - b) \cdot \varphi_C'(b) = \varphi_C(b) - 58 \tag{5.6}$$

Unfortunately, the system of differential equations formed by Equation 5.5 and Equation 5.6 has not an analytical solution, at least to our knowledge. Therefore one has to solve numerically the system. The initial condition requires that

$$\varphi_S(58) = \varphi_C(58) = 58 \tag{5.7}$$

---

[5]To calculate this second probability correctly, it necessary that the coalition use a mechanism which is incentive compatible. Marshall et al. (1994) do not pose a collusive mechanism in the paper. They assume that the values of the members of a coalition are common knowledge within the coalition.

and the end condition requires the existence of a number $b^* \in (58, 100)$ such that

$$\varphi_S(b^*) = \varphi_C(b^*) = 100 \tag{5.8}$$

Let l_i denote the right derivative of φ_i at $b = 58$. Applying l'Hospital rule to Equations 5.5 and 5.6 one has

$$l_C = 2, \quad l_S = \frac{3}{2} \tag{5.9}$$

To solve the problem numerically, it is convenient to reformulate Equations 5.5 and 5.6 in the following way. Let us define

$$\delta_i(b) = \frac{\varphi_i(b) - 58}{b - 58}$$

Applying this transformations to Equations 5.5 and 5.6 we get

$$(\delta_C(b) - 1) \cdot [\delta_S'(b) + (b - 58)^{-1} \delta_S(b)] = \delta_S(b) \tag{5.10}$$
$$2 \cdot (\delta_S(b) - 1) \cdot [\delta_C'(b) + (b - 58)^{-1} \delta_C(b)] = \delta_C(b) \tag{5.11}$$

with the following initial and end conditions

$$\delta_i(58) = l_i, \qquad \delta_i(b^*) = \frac{42}{b^* - 58}, \qquad i = C, S \tag{5.12}$$

Following Marshall et al. (1994), we use Taylor expansions to approximate numerically Equations 5.10 and 5.11. Figure 5.1 shows the equilibrium bid functions for the coalition and for the single player. Note that the coalition shades its bid more than the single player. This is consistent with the result of Maskin and Riley (1995).

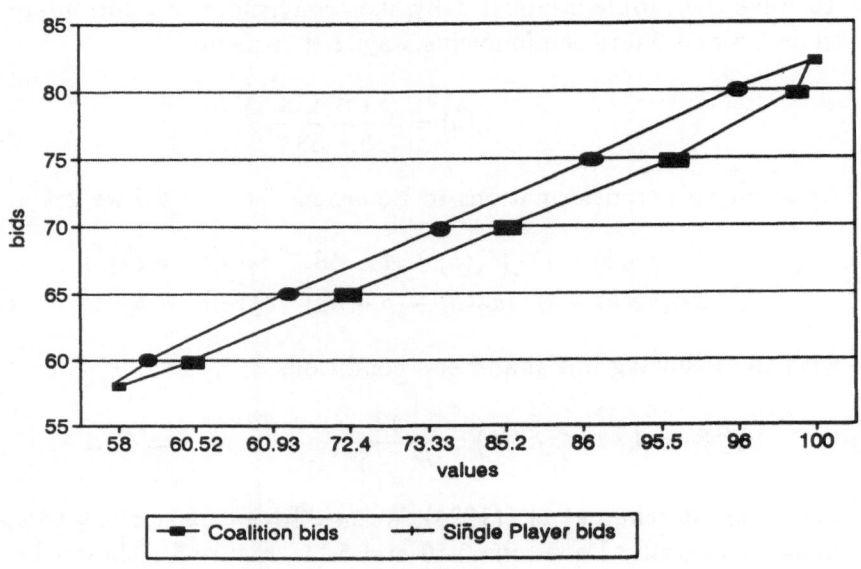

Figure 5.1: Equilibrium bids

Chapter 6

Conclusion

We have opened this work asking the question: "Which mechanisms, if any, are likely to be observed when human subjects have to solve a given implementation problem?". The implementation problem we have investigated is that which is faced by a ring of bidders who participate in a first-price auction. In 92% of all rounds, players agreed on some collusive mechanism and implemented it. The mechanisms used by human players are very simple. Playing according to the most frequently used mechanism, they announced their values. In eleven auctions, players did not care about the announcing order; in one auction, they used a random mechanism to choose in which order to announce; in one auction, they specified the sequence of announcements so that Player i ($i=1,2,3$) once announced as first one, once as second one and so on. The player who made the highest announcement bid the reserve price and splitted equally the difference between his announcement minus the reserve price among the other two players. This mechanism is not incentive-compatible but subjects, in most of the rounds, announced their true values. Because of this, experimental subjects reached the optimal allocation in most of the cases.

Other three mechanisms were observed. The bid-bargain mechanism was used by experimental subjects in two auctions entirely and for ten rounds in Auction 9. Players who played according to this mechanism bargained making bids and asks. The player who made the highest bid bid the reserve price in the legitimate auction and paid each of the losers what he bid them or they asked him, respectively. The lattice mechanism was used only in one auction. Players played in the main auction according to the lattice shown in Table 2.3 in order to choose the winner. They agreed that the winner of the legitimate auction had to make a side payment equal to his value minus the submitted bid divided by three. The first-price auction mechanism was used only in three rounds. Players used a first-price auction to choose the winner. The player who made the highest bid bid the reserve price in the legitimate auction and splitted equally the difference between his bid (in the prior auction) and the reserve price.

Each of these mechanisms has been strategically analyzed. We have found that none of the mechanisms we have observed is incentive-compatible. Players who used the lattice mechanism submitted always their bids according to the lattice shown in Table 2.3 and made their side payments according to the lattice too. Players who used the first-price auction mechanism submitted their own value in all cases but one. Analyzing the bid-bargain mechanism, we have not cared about the asks. That is, we have compared the data with an idealized mechanism in which players make their bids simultaneously. Although this assumption has simplified the analysis, it is difficult to clearly identify features of players' behavior. There is a tendency to overbid the theoretical prediction at low values of the equilibrium bid.

All these mechanisms are non-optimal when they are played according to the theoretical predictions. That is, there is a positive probability that the player with the highest value does not obtain the item. This is partly due to the fact that players have to use integer number. Allowing real numbers, the probability of non-optimality is zero for all mechanisms but the announcement mechanism. For each mechanism, we have calculated the ex ante expected payoff and the ex ante expected loss. They are ranked in Table 4.5.

As we have observed in Chapter 1, to our knowledge, our approach is innovatory. It allows to investigate not only beliefs updating and strategy choice but also the discovery of game mechanisms which players spontaneously construct. This approach also allows us to answer the question whether there are social rules (mechanisms) which have a higher degree of social acceptance than others. Looking at our experimental results, we can conclude that, in the implementation problem we have investigated, the mechanism with the highest degree of social acceptance is the announcement mechanism (which was used in 13 auctions), and it is the one which is the most inefficient when players play according to the theoretical prediction. Since, as we have already observed, experimental subjects announce their true values in most of the rounds, this mechanism provides the optimal allocation in most of the cases.

We want to conclude relating our work to the most recent empirical work on collusion in auctions and to the experimental work on designer markets (Plott (1994)). We also say a few words about the political relevance of our experiment.

Much of the empirical work about collusion in auctions has appeared in the last few years. Most of it concerns either tests to ascertain collusion in auctions or procurements (Howard and Kaserman (1989), Porter and Zona (1993)) or studies that compare the cartel bidding behavior with some theoretical prediction (Pesendorfer (1994b)). The main difficulties of empirical studies are the scarcity of information about the mechanism used and the strategies played by firms to implement collusion. Laboratory experiments allow to analyze players' behavior in an environment in which everything is under control. In this way, laboratory experiments allow also to compare collusion in different kinds of auctions and with a different number of players.

Many experimental investigations on public goods have appeared in the last years[1]. In all this work, one wants to analyze either one given mechanism over a wide range of environments, or a wide range of given mechanisms over a limited class of economic environments. This approach, which also is very promising for policy problems, differs from ours. We have not given any suggestion to experimental subjects about the mechanisms to use. Discovering which mechanisms do experimental subjects use has been the aim of our investigation. On the contrary, the literature on public goods focuses on efficiency questions, that is, on whether one given mechanism is better than another in an implementation problem[2]. Certainly, it would be also interesting to analyze public goods problems with our approach, to compare the mechanisms found out by experimental subjects with those proposed by the literature.

A subject which is strictly related to our work and makes it politically relevant is the collusion in procurement of government contracts. The *Bundeskartellamt*[3] claims that the damages suffered by the taxpayers in Germany in the last 30 years because of collusion in procurement of contracts for the maintenance of the *autobahns* amounts to hundreds of millions of DM (Panorama, ARD (January 1996)). Therefore, it is important to design auctions with rules which do not make collusive behavior easy. One way to test these auctions is to execute laboratory experiments.

Finally, it would be interesting to consider extensions of this work.

---

[1]See Ledyard (1995) for a survey on experimental research on public goods. In Economic Theory (1994, number 1) are published the contributes presented at the symposium "Designer Markets: laboratory experimental methods in economics".

[2]In the public good field, the merger of theory and experimental work is becoming a reality. Smart markets are the product of joint work of theorists with experimenters.

[3]The German antitrust office.

For example, one could analyze the same game with more than three players to see whether players form subcoalitions. One could also allow players to communicate only before they draw their values in order to see whether the effectiveness of nonbinding communication in inducing cooperation is sensitive to the moment in which it happens.

Bibliography

[1] Abreu, D., D. Pearce, and E. Stacchetti, Optimal Cartel Equilibria with Imperfect Monitoring, *Journal of Economic Theory*, 39, 1986, 251-269.

[2] Artale, Angelo (ed. by) Rings in Auctions: An Experimental Approach. Protocols of a Videocamera Experiment (in German), in *Experimental Data Documentation Series* No. 2.96, SFB 303, Bonn Universität, 1996.

[3] Baldwin, Laura H., R. Marshall and J-F. Richard, Bidder Collusion at Forest Service Timber Sales, PennState Working Paper, April 1995.

[4] Bernheim, B.D., B. Peleg, and M.D. Whinston, Coalition-Proof Nash Equilibria I: Concepts, *Journal of Economic Theory*, 42, 1987, 1-12.

[5] Cassady, R., Jr., *Auctions and Auctioneering*, Berkeley: University of California Press, 1967.

[6] Cox, J.C., V.L. Smith and J.M. Walker, Theory and individual behavior of first-price auctions, *Journal of Risk and Uncertainty*, 1, 1988, 61-99.

[7] Dyer, D., J.H. Kagel and D. Levin, Resolving uncertainty about the number of bidders in independent private-value auctions: An experimental analysis, *Rand Journal of Economics*, 20, 1989, 268-279.

[8] Fehl, Ulrich and Werner Güth, Internal and External Stability of Bidder Cartels in Auctions and Public Tenders, *International Journal of Industrial Organization*, 5, 1987, 303-313.

[9] Fudenberg, Drew and Jean Tirole, *Game Theory*, Cambridge, MA: MIT Press, 1991.

[10] Graham, Daniel and Robert C. Marshall, Collusive Bidder Behavior at Single-Price and English Auctions, *Journal of Political Economy*, Dec, 1987, 1217-1239.

[11] Güth, Werner and Bezalel Peleg, On Ring Formation in Auctions, *Mimeo*, Sept. 1993.

[12] Harsanyi, J. Games with Incomplete Information Played by Bayesian Players, *Management Science*, (1967,1968) 14, 159-182, 320-334, 486-502.

[13] Harsanyi, J and R. Selten, A Generalized Nash Solution for Bargaining Games with Incomplete Information, Management Science, Vol. 18, No 5, 1972, 80-106.

[14] Holmström, B and Myerson, R. B., Efficient and Durable Decision Rules with Incomplete Information, *Econometrica* 51, 1983, 1799-1819.

[15] Howard Jeffrey H. and David Kaserman, Proof of damages in construction industry bid-rigging cases, *The Antitrust Bulletin*, Summer 1989, 359-393.

[16] Isaac, Mark R. and M. James Walker, Information and Conspiracy in Sealed Bid Auctions, *Journal of Economic Behavior and Organization*, 6, 1985, 139-159.

[17] Kagel, John H., Auctions: a survey of experimental research, in: *Handbook of experimental economics*, ed. by Kagel, J. and A. Roth. Princeton: Princeton University Press, 1995.

[18] Kreps, D., P. Milgrom, J. Roberts, and R. Wilson, Rational Cooperation in Finitely Repeated Prisoner's Dilemma, *Journal of Economic Theory*, 27, 1982, 245-252.

[19] Kreps, D. and R. Wilson, Sequential Equilibria, *Econometrica*, 50, 1982, 443-459.

[20] Le Brun, Bernard, *Asymmetry in Auctions*, Ph.D dissertation, Catholic University of Louvain, 1991.

[21] Ledyard, John O., Public Goods: A Survey of Experimental Research, in: *Handbook of experimental economics*, ed. by Kagel, J. and A. Roth. Princeton: Princeton University Press, 1995.

[22] Mailath, George and Peter Zemsky, Collusion in Second Price Auctions with Heterogeneous Bidders, *Games and Economic Behavior*, Nov. 1991, 467-86.

[23] Marshall, Robert C., M. Meurer, J-F. Richard and W. Stromquist, Numerical Analysis of Asymmetric First Price Auctions, *Games and Economic Behavior*, 7, 1994, 193-220.

[24] Maskin, Eric and John Riley, Asymmetric Auctions, mimeo Harvard University and UCLA (1995).

[25] Matthews, Steven A., A Technical Primer on Auction Theory, mimeo Northwestern University (1990).

[26] McAfee, Preston and John McMillan, Auctions and Bidding, *Journal of Economic Literature*, 25, 1987, 708-747.

[27] McAfee, Preston and John McMillan, Bidding Rings, *The American Economic Review*, June 1992, 579-599.

[28] Milgrom, R. Paul and R. J. Weber, A Theory of Auctions and Competitive Bidding, *Econometrica*, 50, 1982, 1089-1122.

[29] Milgrom, R. Paul and R.J. Weber, A Theory of Auctions and Competitive Bidding II, 1982, mimeo.

[30] Myerson, Roger and Mark Satterthwaite, Efficient Mechanisms for Bilateral Trading, *Journal of Economic Theory*, 28, 1983, 265-281.

[31] Myerson, Roger B. *Game Theory*, Cambridge, MA, Harvard University Press, 1991.

[32] Palfrey, T., Implementation in Bayesian Equilibrium: The Multiple Equilibrium Problem in Mechanism Design, in J.-J. Laffont (Ed.) *Advances in Economic Theory*, Cambridge: Cambridge University Press, 1992.

[33] Palfrey T. and S. Srivastava, *Bayesian Implementation*, Harwood Academic Publ., 1993.

[34] Pesendorfer, Martin, A Study of Collusion in First Price Auctions Part I: Cartel Agreements, mimeo Northwestern University, 1994a.

[35] Pesendorfer, Martin, A Study of Collusion in First Price Auctions Part II: Cartel Bidding Behavior, mimeo Northwestern University, 1994b.

[36] Plott, Charles R., (edited by) *Symposium Designer Markets: laboratory experimental methods in economics* in *Economic Theory* vol 4, 1, 1994.

[37] Porter, Robert H. and J. Douglas Zona, Detection of Bid Rigging in Procurement Auctions, *Journal of Political Economy*, 101, 1993, 518-538.

[38] Rubinstein, Ariel, Comments on the Interpretation of Game Theory, *Econometrica*, 59, 1991, 909-924.

[39] Siegel, S. and N.J. Castellan, Jr., *Nonparametric Statistics*, 2nd Edition, New York, McGraw-Hill, Inc. 1988.

[40] Selten, Reinhard, Equal Share Analysis of Characteristic Function Experiments, in *Contribution to Experimental Economics, vol. III*, ed. H. Sauermann, Tübingen, 1972.

[41] Selten, Reinhard, Reexamination of the Perfectness Concept for Equilibrium Points in Extensive Games, *International Journal of Game Theory*, 4, 1975, 25-55.

[42] Selten, Reinhard, The Chain Store Paradox, *Theory and Decision*, 9, 1978, 127-159.

[43] Selten, Reinhard, Bounded Rationality, *Journal of Institutional and Theoretical Economics*, 146, 1990 649-658.

[44] Selten, Reinhard, Evolution, Learning, and Economic Behavior, *Games and Economic Behavior*, 3, 1991, 3-24.

[45] Selten, Reinhard and Rolf Stoecker, End Behavior in Sequences of Finite Prisoner's Dilemma Supergames, *Journal of Economic Behavior and Organization*, 7, 1986, 47-70.

[46] Theil, H. *Principles of Econometrics*, New York, John Wiley & Sons, Inc., 1971.

[47] Vickrey, William, Counterspeculation, Auction, and Competitive Sealed Tenders, *Journal of Finance*, 16, 1961, 8-37.

[48] Weber, Robert J., Multiple-Object Auctions, in *Auctions, Bidding, and Contracting*, ed. Richard Engelbrecht-Wiggans, Martin Shubik, and Robert M. Stark, New York, NYU Press, 1983.

[49] Wilson, Robert, Strategic Analysis of Auctions, in *The Handbook of Game Theory*, ed Robert Aumann and Sergiu Hart, Amsterdam, North-Holland, 1992.

Appendix A

Instructions and Forms

In this appendix, we present the instructions as they were read to the experimental subjects. We also reproduce the forms received by experimental subjects to make their bids and side payments, respectively.

Anweisungen

Bei diesem Experiment handelt es sich um eine Auktion. Die Regeln der Auktion werden unten erläutert.

Wiederverkaufswert Ein Objekt wird versteigert. Jeder von Ihnen hat einen **Wiederverkaufswert**, d.h. falls man das Objekt erhält, bekommt man genau diesen Wert bei einem fiktivem Verkauf.

Wiederverkaufswertsbestimmung Die Wiederverkaufswerte werden von jedem einzelnen Teilnehmer durch eine Ziehung von einem Kartenstapel, der 50 Karten, von **51** bis **100**, enthält, **zufällig** bestimmt. Nach jeder Ziehung wird die Karte auf dem Stapel zurückgelegt, der dann wiedergemischt wird. Jeder erfährt nur seinen eigenen Wiederverkaufswert.

Wer ist der Erwerber des Objekts? Um das Objekt erhalten zu können, muß man wenigstens **58, das Mindestgebot**, anbieten. Der **Erwerber** ist derjenige, der das höchste Gebot abgibt. Sein **Gewinn** ist gleich Wiederverkaufswert minus Gebot. Falls zwei oder mehrere Bieter dasselbe höchste Gebot abgeben, wird gewürfelt, um den Erwerber zu bestimmen. Dabei hat jeder Höchstbieter die gleiche Chance. Falls kein Teilnehmer einen höheren Wiederverkaufswert als das Mindestgebot hat, werden die Wiederverkaufswerte erneut festgelegt, bis der Wiederverkaufswert wenigstens für einen Bieter höher als das Mindestgebot ist.

Absprachemöglichkeit Nachdem **alle** Ihre Wiederverkaufswerte erfahren haben, können Sie sich zusammen mit den anderen Teilnehmern an den Tisch setzen, der sich in der Mitte des Raums befindet. Sie dürfen miteinander reden und sich Notizen machen.

Abgabe der Gebote Wenn wenigstens zwei Teilnehmer fertig sind, sollen **alle gleichzeitig** Ihr Gebot auf den Vordruck schreiben, der auf Ihrem eigenem Tisch liegt. Das Gebot jedes Teilnehmers muß immer **kleiner oder gleich 100** sein.

Mitteilung des Erwerbers und seines Gebots Nachdem Sie Ihr Gebot gemacht haben, werden der Erwerber und sein Gebot bekanntgegeben.

Transfersmöglichkeit Es ist dem Erwerber gestattet, Geld an die anderen Teilnehmer durch ein Formular zu überweisen. In jeder Runde muß die **gesamte Geldüberweisung** (überwiesenes Geld an beide anderen Teilnehmer) **kleiner** als der in jeder Runde realisierte Gewinn sein. Der Betrag der Geldüberweisung an jeden Teilnehmer wird allen bekanntgegeben.

Unsere Währung Die Gewinne werden in *Soldi* ausgerechnet, die am Ende in DM umgerechnet werden. Der Wechselkurs ist $\frac{DM}{Soldo} = 0.20$

Die Auktion wird wiederholt Die Auktion wird 20 Male wiederholt. Vor jeder Runde ziehen Sie Ihre neuen Wiederverkaufswerte. Das Mindestgebot bleibt unverändert.

Es ist nicht erlaubt mehr als 10 *Soldi* zu verlieren Am Anfang der Auktion bekommt jeder von Ihnen eine Erstausstattung von 10 *Soldi*. Wenn ein Teilnehmer im Laufe des Spiels mehr als 10 *Soldi* verloren hat, muß er die Auktion verlassen.

Noch etwas Während Sie miteinander sprechen, werden Sie gefilmt. Außerdem werden Sie in einigen Runden zum Erklären Ihrer Entscheidung aufgefordert.

Viel Erfolg!!!

Rheinische Friedrich-Wilhelms-Universität Bonn
Institut für Gesellschafts- und Wirtschaftswissenschaften

Experimentelle Untersuchung

Spieler 1   Runde 1   au2610n

Schreiben Sie hier bitte Ihr Gebot: *Soldi*___

Erklären Sie bitte Ihre Entscheidung

Rheinische Friedrich-Wilhelms-Universität Bonn
Institut für Gesellschafts- und Wirtschaftswissenschaften

Experimentelle Untersuchung

Überweisungsformular
au2610n

1. Der Spieler .... überweist

   ....Soldi dem Spieler 1

   ....Soldi dem Spieler 2

   ....Soldi dem Spieler 3

   Unterschrift

2. Der Spieler .... überweist

   ....Soldi dem Spieler 1

   ....Soldi dem Spieler 2

   ....Soldi dem Spieler 3

   Unterschrift

3. Der Spieler .... überweist

   ....Soldi dem Spieler 1

   ....Soldi dem Spieler 2

   ....Soldi dem Spieler 3

   Unterschrift

Appendix B

Experimental Data

In this appendix, we want to present some experimental results more in detail. For each auction, the first column $i_{(1)}$ represents the player with the highest value, and who would be selected by an optimal mechanism. The second column $v_{(1)}$ represents the highest drawn value. The third column w represents the player who won the round. b_w is the bid submitted by the winner. π_w and π_l are the winner's and losers' payoffs, respectively. Since sometimes the payoffs of the two losers are not the same, we adopt the following convention: the payoff on the left side is always the payoff of the player with the smallest number. For example, if Player 2 is the winner, $x - y$ under π_l means that x is the payoff of Player 1 and y is the payoff of Player 3. p_w and p_l are the winner's and losers' payoffs in the case that subjects play the mechanism proposed by McAfee and McMillan (1992). Note that adapting the solution proposed by McAfee and McMillan to our problem, we just round off $T(v)$, as defined in Section 2.3, disregarding the units problem. KK^* is the ex post coefficient of cooperation as defined in Subsection 2.4.1 (see Artale (1996) for details).

Notation
$i_{(1)}$ = Player with highest value
$v_{(1)}$ = Highest value
w = Winner
b_w = Winner's bid
π_w = Winner's payoff
π_l = Losers' payoffs
p_w = Winner's payoff according to McAfee and McMillan's theory
p_l = Losers' payoff according to McAfee and McMillan's theory
KK^* = Ex post cooperation (Y=yes, N=no)

$(R) =$ The winner was randomly chosen, since two or all bidders submitted the same bid

$w() =$ It specifies the winner's value when it does not coincide with $v_{(1)}$

RA=random announcement mechanism

SA=simple announcement mechanism

FO=fixed order announcement mechanism

BB=bid-bargain mechanism

LA=lattice mechanism

FPA=first-price auction

Auction 1

$i_{(1)}$	$v_{(1)}$	w	b_w	π_w	π_l	p_w	p_l	KK^*	M
3	72	3	58	6	4-4	8	3	Y	RA
2,3	74	2	58	6	5-5	9	3,5	Y	RA
2	86	2	58	14	7-7	14	7	Y	RA
3	100	3	60	20	10-10	21	10,5	Y	RA
2	97	2	58	15	12-12	20	9,5	Y	RA
1	93	1	58	13	11-11	18	8,5	Y	RA
2	78	2	58	7	7-6	11	4,5	Y	RA
2	80	2	58	9	6-7	12	5	Y	RA
2	99	2	58	18	11-12	21	10	Y	RA
1,3	74	1	58	6	5-5	9	3,5	Y	RA
2	90	2	58	12	10-10	16	8	Y	RA
2,3	100	3	59	13	14-14	21	10,5	Y	RA
2	64	2	58	2	2-2	4	1	Y	RA
3	93	3	58	12	12-11	18	8,5	Y	RA
3	86	3	58	14	7-7	14	7	Y	RA
1	88	1	58	10	10-10	15	7,5	Y	RA
1	80	1	58	8	7-7	12	5	Y	RA
3	87	3	58	10	10-9	15	7	Y	FPA
2	96	2	58	13	12-13	19	9,5	Y	FPA
3	95	3	58	3	17-17	19	9	Y	FPA

Auction 2

$i_{(1)}$	$v_{(1)}$	w	b_w	π_w	π_l	p_w	p_l	KK^*	**M**
3	98	3	59	19	10-10	20	10	Y	BB
2	100	2	58	20	12-10	21	10,5	Y	BB
3	87	3	58	13	10-6	15	7	Y	BB
3	80	3	58	11	7-4	12	5	Y	BB
1,2	100	2	58	14	14-14	21	10,5	Y	BB
3	69	3	58	5	3-3	6	2,5	Y	BB
2	100	2	58	16	10-16	21	10,5	Y	BB
2	97	2	58	19	10-10	20	9,5	Y	BB
3	94	3	58	17	9-10	18	9	Y	BB
3	73	3	58	5	5-5	8	3,5	Y	BB
2	87	2	58	19	5-5	15	7	Y	BB
1	61	1	58	1	1-1	2	0,5	Y	BB
1	96	1	58	18	10-10	19	9,5	Y	BB
2	100	2	58	22	10-10	21	10,5	Y	BB
1	87	1	58	11	9-9	15	7	Y	BB
2	91	2	58	13	11-9	17	8	Y	BB
1	100	1	58	25	8-9	21	10,5	Y	BB
3	86	3	58	13	8-7	15	6,5	Y	BB
1	95	1	59	36	0-0	19	9	N	BB
1	90	1	58	12	10-10	17	7,5	Y	BB

Auction 3

$i_{(1)}$	$v_{(1)}$	w	b_w	π_w	π_l	p_w	p_l	KK^*	M
1	100	1	58	14	14-14	21	10,5	Y	SA
3	91	3	58	11	11-11	17	8	Y	SA
3	98	3	58	14	13-13	20	10	Y	SA
1	94	1	58	12	12-12	18	9	Y	SA
1	94	1	58	12	12-12	18	9	Y	SA
2	89	2	58	11	10-10	16	7,5	Y	SA
3	88	3	58	10	10-10	16	7	Y	SA
1	68	1	58	4	3-3	6	2	Y	SA
1	92	1	58	12	11-11	18	8	Y	SA
2	93	2	58	13	11-11	18	8,5	Y	SA
1	80	1	58	2	10-10	12	5	Y	SA
2	80	2	58	8	7-7	12	5	Y	SA
3	80	3	58	8	7-7	12	5	Y	SA
1	91	1	58	15	9-9	17	8	Y	SA
1	91	1	58	11	11-11	17	8	Y	SA
2	97	2	58	13	13-13	20	9,5	Y	SA
3	87	3	58	11	9-9	15	7	Y	SA
2	98	1	59	38	0-0	20	10	N	SA
2	72	2	64	4	0-4	-	-	N	SA (2 and 3)
3	83	3	70	7	0-6	-	-	N	SA (2 and 3)

Auction 4

$i_{(1)}$	$v_{(1)}$	w	b_w	π_w	π_l	p_w	p_l	KK^*	M
1	90	1	59	11	10-10	17	7,5	Y	SA
1	100	1	59	15	13-13	21	10,5	Y	SA
1	83	1	59	8	8-8	13	6	Y	SA
1	98	1	58	20	10-10	20	10	Y	SA
2	87	2	58	11	9-9	15	7	Y	SA
3	87	3	58	11	9-9	15	7	Y	SA
1	73	1	58	5	5-5	8	3,5	Y	SA
3	92	3	58	15	5-14	18	8	Y	SA
1	100	1	58	14	14-14	21	10,5	Y	SA
3	78	3	58	8	6-6	11	4,5	Y	SA
2	96	2	58	16	11-11	19	9,5	Y	SA
2	96	2	58	16	11-11	19	9,5	Y	SA
1	85	1	58	9	9-9	14	6,5	Y	SA
3	85	3	58	9	9-9	14	6,5	Y	SA
2	92	2	58	12	11-11	18	8	Y	SA
2	76	2	58	6	6-6	10	4	Y	SA
1	75	1	58	7	5-5	9	4	Y	SA
3	98	3	58	14	13-13	20	10	Y	SA
3	68	3	58	4	3-3	6	2	Y	SA
2	92	2	58	14	10-10	18	8	Y	SA

Auction 5

$i_{(1)}$	$v_{(1)}$	w	b_w	π_w	π_l	p_w	p_l	KK^*	M
2	81	2	58	8	7-8	12	5,5	Y	SA
2	100	2	58	14	14-14	21	10,5	Y	SA
3	83	3	58	8	8-9	13	6	Y	SA
3	85	2(81)	58	7	8-8	14	6,5	Y	SA
2	81	2	58	8	7-8	12	5,5	Y	SA
3	93	3	58	17	9-9	18	8,5	Y	SA
3	95	3	58	21	8-8	19	9	Y	SA
3	93	1(92)	58	12	11-11	18	8,5	Y	SA
1	94	1	58	14	11-11	18	9	Y	SA
2	67	2	58	3	3-3	5	2	Y	SA
1	99	1	58	17	12-12	21	10	Y	SA
3	73	3	58	8	3-4	8	3,5	Y	SA
3	97	2(93)	58	12	11-12	20	9,5	Y	SA
3	99	3	58	34	4-3	21	10	Y	SA
1	100	1	58	14	14-14	21	10,5	Y	SA
1	65	1	58	3	3-1	4	1,5	Y	SA
2	83	2	58	9	8-8	13	6	Y	SA
3	98	3	58	40	0-0	20	10	N	SA
3	92	3	75	17	0-0	18	8	N	No Mechanism
3	91	3	75	16	0-0	17	8	N	No Mechanism

Auction 6

$i_{(1)}$	$v_{(1)}$	w	b_w	π_w	π_l	p_w	p_l	KK^*	M
1	93	1	83	10	0-0	18	8,5	N	BB
1	100	1	60	16	12-12	21	10,5	Y	BB
1	99	3(92)	88	4	0-0	21	10	N	BB
2	82	3(78)	76	2	0-0	13	5,5	N	BB
3	88	3	59	9	10-10	16	7	Y	BB
1	91	1	58	15	8-10	17	8	Y	BB
2	74	1(53)	58	-5	0-0	9	3,5	Y	BB
3	89	3	58	31	0-0[1]	16	7,5	Y	BB
3	79	2(74)	58	16	0-0[2]	11	5	Y	BB
1	92	1	58	14	10-10	18	8	Y	BB
2	81	2	58	9	7-7	12	5,5	Y	BB
3	95	3	58	18	12-7	19	9	Y	BB
2	89	2	58	13	9-9	16	7,5	Y	BB
3	93	3	58	16	8-11	18	8,5	Y	BB
2	86	2	58	12	8-8	15	6,5	Y	BB
1	70	1	58	6	3-3	7	2,5	Y	BB
1	73	1	65	8	0-0	8	3,5	N	BB
3	95	3	58	20	9-8	19	9	Y	BB
1	95	1	58	19	9-9	19	9	Y	BB
1	81	1	71	10	0-0	12	5,5	N	BB

Auction 7

$i_{(1)}$	$v_{(1)}$	w	b_w	π_w	π_l	p_w	p_l	KK^*	M
1	99	3(R)	61	12	12-12	21	10	Y	LA
1	95	1	61	12	11-11	19	9	Y	LA
2	95	2	61	12	11-11	19	9	Y	LA
2	78	2(R)	58	8	6-6	11	4,5	Y	LA
3	89	3	60	9	10-10	16	7,5	Y	LA
1	90	1	60	10	10-10	17	7,5	Y	LA
1	89	1(R)	60	9	10-10	16	7,5	Y	LA
3	97	3	61	12	12-12	20	9,5	Y	LA
3	96	3	61	13	11-11	19	9,5	Y	LA
3	99	3	61	14	12-12	21	10	Y	LA
2	69	1(R)	58	0	2-2	6	2,5	Y	LA
2	88	2	60	10	9-9	16	7	Y	LA
3	92	3	61	11	10-10	18	8	Y	LA
1	91	1	61	8	11-11	17	8	Y	LA
1	83	1	60	7	8-8	13	6	Y	LA
2	95	2	61	12	11-11	19	9	Y	LA
2	88	2	60	10	9-9	16	7	Y	LA
1	83	1	60	7	8-8	13	6	Y	LA
2	99	2(R)	61	13	12-12	21	10	Y	LA
2	85	2	60	9	8-8	14	6,5	Y	LA

Auction 8

$i_{(1)}$	$v_{(1)}$	w	b_w	π_w	π_l	p_w	p_l	KK^*	**M**
1	91	1	58	11	11-11	17	8	Y	SA
1	97	1	58	13	13-13	20	9,5	Y	SA
2	89	2	58	11	10-10	16	7,5	Y	SA
2	80	2	58	8	7-7	12	5	Y	SA
2	92	2	58	12	11-11	18	8	Y	SA
1	98	1	58	14	13-13	20	10	Y	SA
3	78	3	58	8	6-6	11	4,5	Y	SA
1,2,3	60	3(R)	58	2	0-0	1	0,5	Y	SA
1	91	1	58	11	11-11	17	8	Y	SA
1	82	1	58	8	8-8	13	5,5	Y	SA
3	88	3	58	10	10-10	16	7	Y	SA
2	88	2	58	10	10-10	16	7	Y	SA
3	76	3	58	6	6-6	10	4	Y	SA
1,3	89	3	58	11	10-10	16	7,5	Y	SA
2	87	2	58	11	9-9	15	7	Y	SA
2	92	2	58	14	10-10	18	8	Y	SA
3	82	3	58	8	8-8	13	5,5	Y	SA
1	65	1	58	3	2-2	4	1,5	Y	SA
3	95	3	58	13	12-12	19	9	Y	SA
1	99	1	58	15	13-13	21	10	Y	SA

Auction 9

$i_{(1)}$	$v_{(1)}$	w	b_w	π_w	π_l	p_w	p_l	KK^*	M
2	85	2	58	27	0-0	14	6.5	Y	SA
2	85	2	58	9	9-9	14	6.5	Y	SA
3	79	3	58	11	10-0	11	5	Y	SA
2	100	2	58	3	20-19	21	10.5	Y	SA
1	79	1	58	7	7-7	11	5	Y	SA
2	90	2	58	10	11-11	17	7.5	Y	SA
1	97	1	58	13	13-13	20	9.5	Y	SA
3	88	3	58	10	10-10	16	7	Y	SA
3	74	3	58	5	5-6	9	3.5	Y	SA
1	100	1	58	14	14-14	21	10.5	Y	SA
3	71	3	58	5	4-4	7	3	Y	SA
3	77	1(R)	60	15	0-0	10	4.5	N	BB
2,3	100	3	70	30	0-0	21	10.5	N	BB
1	93	1	70	23	0-0	18	8.5	N	BB
3	99	3	58	23	9-9	21	10	Y	BB
2	93	2	58	15	10-10	18	8.5	Y	BB
1	94	1	58	22	7-7	18	9	Y	BB
3	96	3	58	24	7-7	19	9.5	Y	BB
3	100	3	58	24	10-8	21	10.5	Y	BB
1	79	1	58	7	7-7	11	5	Y	SA

Auction 10

$i_{(1)}$	$v_{(1)}$	w	b_w	π_w	π_l	p_w	p_l	KK^*	M
1	98	1	59	13	13-13	20	10	Y	SA
1,2	78	1(R)	58	8	6-6	11	4.5	Y	SA
1,3	76	3(R)	58	6	6-6	10	4	Y	SA
1	82	1	59	7	8-8	13	5.5	Y	SA
2	97	2	58	13	13-13	20	9.5	Y	SA
1	99	1	58	15	13-13	21	10	Y	SA
1	77	1	58	7	6-6	10	4.5	Y	SA
3	77	3	58	7	6-6	10	4.5	Y	SA
2	85	2	58	9	9-9	14	6.5	Y	SA
2	89	2	58	11	10-10	16	7.5	Y	SA
3	71	3	58	5	4-4	7	3	Y	SA
1	78	1	58	6	7-7	11	4.5	Y	SA
3	82	3	58	8	8-8	13	5.5	Y	SA
3	79	3	58	7	7-7	11	5	Y	SA
2	88	2	58	10	10-10	16	7	Y	SA
3	96	3	58	12	13-13	19	9.5	Y	SA
2	90	2	58	10	11-11	17	7.5	Y	SA
2	78	2	58	8	6-6	11	4.5	Y	SA
1	91	1	58	11	11-11	17	8	Y	SA
2	97	2	59	12	13-13	20	9.5	Y	SA

Auction 11

$i_{(1)}$	$v_{(1)}$	w	b_w	π_w	π_l	p_w	p_l	KK^*	M
2	76	2	58	6	6-6	10	4	Y	FA
1	100	1	58	20	11-11	21	10.5	Y	FA
1	87	1	58	11	9-9	15	7	Y	FA
1	85	1	58	9	9-9	14	6.5	Y	FA
1	99	1	58	23	9-9	21	10	Y	FA
1	84	1	58	10	8-8	14	6	Y	FA
1	89	1	58	31	0-0	16	7.5	N	FA
3	88	1(84)	59	15	5-5	16	7	N	FA
3	78	3	59	8	1-10	11	4.5	Y	FA
2	97	2	60	11	13-13	20	9.5	Y	FA
1	83	1	58	17	4-4	13	6	Y	FA
3	97	3	58	17	11-11	20	9.5	Y	FA
2	71	2	58	5	4-4	7	3	Y	FA
3	94	3	58	16	10-10	18	9	Y	FA
1	93	1	60	17	8-8	18	8.5	Y	FA
1	74	2(65)	59	0	3-3	9	3.5	N	FA
2	82	2	61	21	0-0	13	5.5	N	FA
3	91	3	70	18	3-0	17	8	N	FA
1	86	2(77)	72	5	0-0	15	6.5	N	FA
3	90	3	79	11	0-0	17	7.5	N	FA

Auction 12

$i_{(1)}$	$v_{(1)}$	w	b_w	π_w	π_l	p_w	p_l	KK^*	$\mathbf{M}$
3	92	3	58	12	11-11	18	8	Y	SA
2	88	2	58	10	10-10	16	7	Y	SA
3	98	2(92)	58	12	11-11	20	10	Y	SA
3	82	3	58	8	8-8	13	5.5	Y	SA
2	82	2	58	8	8-8	13	5.5	Y	SA
1	94	1	58	14	11-11	18	9	Y	SA
2	75	2	58	7	5-5	9	4	Y	SA
2	78	2	58	8	6-6	11	4.5	Y	SA
2	90	2	58	10	11-11	17	7.5	Y	SA
2	85	2	58	9	9-9	14	6.5	Y	SA
2,3	74	3	58	6	5-5	9	3.5	Y	SA
1	97	2(96)	58	14	12-12	20	9.5	Y	SA
1	86	1	58	12	8-8	15	6.5	Y	SA
2	91	2	58	11	11-11	17	8	Y	SA
2	79	2	58	7	7-7	11	5	Y	SA
2	86	2	58	8	10-10	15	6.5	Y	SA
3	93	3	58	15	10-10	18	8.5	Y	SA
1	94	1	58	14	11-11	18	9	Y	SA
2	91	2	58	11	11-11	17	8	Y	SA
3	82	3	58	14	5-5	13	5.5	Y	SA

Auction 13

$i_{(1)}$	$v_{(1)}$	w	b_w	π_w	π_l	p_w	p_l	KK^*	M
3	94	3	58	12	12-12	18	9	Y	SA
1	93	1	58	13	11-11	18	8.5	Y	SA
3	81	3	58	9	7-7	12	5.5	Y	SA
1	95	1	58	13	12-12	19	9	Y	SA
1	93	1	58	13	11-11	18	8.5	Y	SA
1	75	1	58	7	5-5	9	4	Y	SA
2	97	2	58	13	13-13	20	9.5	Y	SA
1	81	1	58	9	7-7	12	5.5	Y	SA
1	79	1	58	7	7-7	11	5	Y	SA
2	88	2	58	18	6-6	16	7	Y	SA
3	84	3	58	10	8-8	14	6	Y	SA
1	92	1	58	14	10-10	18	8	Y	SA
2	90	2	58	18	7-7	17	7.5	Y	SA
2	59	2	58	1	0-0	1	0	Y	SA
3	93	3	58	13	11-11	18	8.5	Y	SA
3	89	3	58	11	10-10	16	7.5	Y	SA
2	98	2	58	24	8-8	20	10	Y	SA
2	82	2	58	10	7-7	13	5.5	Y	SA
1	94	1	58	16	10-10	18	9	Y	SA
2	93	2	58	19	8-8	18	8.5	Y	SA

Auction 14

$i_{(1)}$	$v_{(1)}$	w	b_w	π_w	π_l	p_w	p_l	KK^*	M
1	82	1	59	9	7-7	13	5.5	Y	SA
3	100	3	58	14	14-14	21	10.5	Y	SA
2	92	2	58	12	11-11	18	8	Y	SA
2	100	2	58	14	14-14	21	10.5	Y	SA
1	80	1	58	8	7-7	12	5	Y	SA
2	100	2	58	14	14-14	21	10.5	Y	SA
1	66	1	58	2	3-3	5	1.5	Y	SA
2	91	2	58	11	11-11	17	8	Y	SA
2	94	2	58	12	12-12	18	9	Y	SA
2	68	2	58	4	3-3	6	2	Y	SA
3	94	3	58	12	12-12	18	9	Y	SA
1	78	1	58	8	6-6	11	4.5	Y	SA
1	98	1	58	14	13-13	20	10	Y	SA
3	89	3	58	11	10-10	16	7.5	Y	SA
2	91	2	58	11	11-11	17	8	Y	SA
1	99	1	58	13	14-14	21	10	Y	SA
3	99	3	58	13	14-14	21	10	Y	SA
2,3	85	2	58	9	9-9	14	6.5	Y	SA
3	99	3	58	13	14-14	21	10	Y	SA
3	71	3	58	5	4-4	7	3	Y	SA

Auction 15

$i_{(1)}$	$v_{(1)}$	w	b_w	π_w	π_l	p_w	p_l	KK^*	M
3	98	3	58	16	12-12	20	10	Y	SA
1	92	3(89)	58	11	10-10	18	8	Y	SA
2	80	2	58	8	7-7	12	5	Y	SA
2	77	1(66)	58	4	2-2	10	4.5	Y	SA
2	99	2	59	14	13-13	21	10	Y	SA
2	86	2	58	10	9-9	15	6.5	Y	SA
2	82	2	58	10	7-7	13	5.5	Y	SA
3	86	3	58	9	9-10	15	6.5	Y	SA
1	96	1	58	14	12-12	19	9.5	Y	SA
2	79	2	58	7	7-7	11	5	Y	SA
1	97	1	58	13	13-13	20	9.5	Y	SA
1	100	1	58	14	14-14	21	10.5	Y	SA
1	80	1	58	8	7-7	12	5	Y	SA
3	98	3	58	13	14-13	20	10	Y	SA
2	62	2	58	2	1-1	3	0.5	Y	SA
2	95	2	58	17	10-10	19	9	Y	SA
2	75	2	58	7	5-5	9	4	Y	SA
3	83	2(82)	58	8	8-8	13	6	Y	SA
1	90	3(R)	70	-7	0-0	17	7.5	N	No Mechanism
2	96	2	90	6	0-0	19	9.5	N	No Mechanism

Auction 16

$i_{(1)}$	$v_{(1)}$	w	b_w	π_w	π_l	p_w	p_l	KK^*	M
1	99	1	58	15	13-13	22	9.5	Y	SA
1	100	1	58	14	14-14	22	10	Y	SA
3	66	3	58	4	2-2	5	1.5	Y	SA
2	81	2	58	9	7-7	13	5	Y	SA
3	76	3	58	6	6-6	10	4	Y	SA
2	99	2	58	15	13-13	22	9.5	Y	SA
3	77	3	58	7	6-6	11	4	Y	SA
2	90	2	58	12	10-10	17	7.5	Y	SA
2	100	2	58	14	14-14	22	10	Y	SA
1	97	1	58	13	13-13	21	9	Y	SA
1	99	1	58	15	13-13	22	9.5	Y	SA
2	97	2	58	13	13-13	21	9	Y	SA
1	86	1	58	10	9-9	15	6.5	Y	SA
2	91	2	58	11	11-11	18	7.5	Y	SA
3	92	3	58	12	11-11	18	8	Y	SA
3	90	3	58	12	10-10	17	7.5	Y	SA
1	99	1	58	15	13-13	22	9.5	Y	SA
1-2	75	2	58	6	6-5	10	3.5	Y	SA
2	100	1(98)	59	39	0-0	21	9.5	N	SA
3	100	3	85					N	No Mechanism

Appendix C

Experimental Data

In this appendix, we present all data which are necessary to make the analysis presented in Section 2.5. A legend, at the end of each table, explains the notation. The theoretical predictions given in what follows are provided in Section 4.2 for the announcement mechanism, in Section 4.3 for the bid-bargain mechanism, in Section 4.4 for the lattice mechanism and in in Section 4.5 for the first-price mechanism, respectively. To indicate which mechanism is used in each round, we use the following abbreviations

RA=random announcement mechanism
SA=simple announcement mechanism
FO=fixed order announcement mechanism
BB=bid-bargain mechanism
LA=lattice mechanism
FPA=first-price auction

Auction 1 (1-17) RA

Position 1				Position 2				Position 3			
i	v	$\hat{v}$	v^*	i	v	$\hat{v}$	v^*	i	v	$\hat{v}$	v^*
2	60	60	63	1	54	54	54	3	72	68	63
2	74	74	75	1	69	69	69	3	74	74	75
1	69	69	72	3	75	72	72	2	86	80	75
1	87	87	84	3	100	90	90	2	60	60	60
1	56	56	56	2	97	94	81	3	56	56	56
1	93	93	87	3	69	68	69	2	60	60	60
2	78	78	78	3	69	69	69	1	69	69	69
1	60	60	63	3	62	62	66	2	80	77	63
3	70	70	72	2	99	93	84	1	60	60	60
2	60	60	63	1	74	74	72	3	74	74	75
1	80	80	81	3	56	56	56	2	90	89	81
3	100	94	90	1	79	79	79	2	100	100	95
3	60	60	63	1	60	60	63	2	64	64	63
1	92	92	87	2	84	84	84	3	93	93	93
2	63	63	66	1	77	77	72	3	86	80	78
2	71	73	72	1	88	88	78	3	80	80	80
2	69	69	72	3	62	62	62	1	80	80	72

LEGEND

i: Player's number

v: Value

$\hat{v}$: Observed announcement

v^*: Equilibrium announcement (given other players' announcements)

Auction 1 (18-20) FPA

Player 1			Player 2			Player 3		
v	g	g^*	v	g	g^*	v	g	g^*
66	66	62	84	80	76	87	87	78
85	85	76	96	96	85	55	55	55
72	72	67	85	85	76	95	95	84

LEGEND:

v: Value

g: Submitted bid

g^*: Equilibrium bid

Auction 2 BB

Player 1			Player 2			Player 3		
v	$\hat{b}$	b^*	v	$\hat{b}$	b^*	v	$\hat{b}$	b^*
69	-	4	61	-	1	98	10	5
60	0	1	100	11	2	52	-10	0
86	8	9	60	0	1	87	8	10
60	-	1	57	2	0	80	6	2
100	14	14	100	14	14	98	13	13
54	2	0	53	-	0	69	3	1
78	5	7	100	13	13	99	13	13
83	7	8	97	10	9	70	-	6
65	-	3	87	9	10	94	10	11
57	0	0	66	-	4	73	5	5
66	2	4	87	5	5	52	-4	0
61	1	1	51	0	0	59	-	0
96	10	10	86	9	9	77	-9	7
67	3	4	100	10	9	82	8	8
87	9	10	84	8	9	72	-	6
89	10	10	91	10	11	60	-9	1
100	8	1	52	-8	0	55	-9	0
71	4	6	75	5	7	86	8	8
95			93			64		
90			88			51		

LEGEND

v: Value

$\hat{b} \geq 0$: Highest observed bid

$\hat{b} < 0$: Lowest observed ask

b^*: Equilibrium highest bid

-: Neither bid nor ask were made

Auction 3 SA

Position 1				Position 2				Position 3			
i	v	$\hat{v}$	v^*	i	v	$\hat{v}$	v^*	i	v	$\hat{v}$	v^*
1	100	100	90	2	62	62	62	3	64	64	64
3	91	91	87	2	70	70	70	1	69	69	69
3	98	98	90	2	52	52	52	1	60	60	60
1	94	94	87	3	78	78	78	2	79	79	79
2	92	92	87	1	94	94	93	3	52	52	52
3	70	70	72	2	89	89	78	1	53	53	53
2	60	60	63	3	88	88	78	1	76	76	76
3	55	55	55	2	63	63	66	1	68	68	66
2	66	66	69	1	92	92	81	3	56	56	56
1	91	91	87	3	61	61	61	2	93	93	93
1	80	80	81	3	67	67	67	2	67	67	67
3	75	75	75	2	80	80	78	1	68	68	68
1	73	73	75	2	56	56	56	3	80	80	75
1	91	91	87	2	85	85	85	3	65	65	65
3	63	63	66	2	52	52	52	1	91	91	66
2	97	97	90	3	88	88	88	1	93	93	93
2	67	67	72	1	84	84	75	3	87	87	87
2	98	98	90	1	97	97	97	3	69	69	69
1	51	-	-	2	72	-	-	3	66	-	-
1	76	-	-	2	53	-	-	3	83	-	-

LEGEND

i: Player's number

v: Value

$\hat{v}$: Observed announcement

v^*: Equilibrium announcement (given other players' announcements)

In Round 18, Player 2 defected. He did not make any side payment.

Auction 4 SA

Position 1				Position 2				Position 3			
i	v	$\hat{v}$	v^*	i	v	$\hat{v}$	v^*	i	v	$\hat{v}$	v^*
2	69	69	72	1	90	90	78	3	87	87	87
3	60	60	63	2	62	62	66	1	100	100	63
2	63	63	66	3	54	54	54	1	83	83	66
3	82	82	81	1	98	98	84	2	87	87	87
2	87	87	84	1	72	72	72	3	53	53	53
2	59	59	60	3	87	87	78	1	54	54	54
3	51	51	51	2	65	65	66	1	73	73	66
2	77	77	78	1	73	73	73	3	92	92	78
2	64	64	69	3	51	51	51	1	100	100	66
3	78	78	78	2	64	64	64	1	60	60	60
3	90	90	87	2	96	92	93	1	64	64	64
2	96	92	90	3	74	74	74	1	69	69	69
2	52	52	52	3	60	60	63	1	85	85	63
3	85	85	84	2	83	83	83	1	64	64	64
1	57	57	57	3	55	55	60	2	91	91	60
2	76	76	78	1	57	57	57	3	55	55	55
2	61	61	63	3	56	56	56	1	75	74	63
2	80	78	81	1	98	98	84	3	82	82	82
2	55	55	55	1	56	56	60	3	68	68	60
3	56	56	56	2	92	90	81	1	58	58	58

LEGEND

i: Player's number

v: Value

$\hat{v}$: Observed announcement

v^*: Equilibrium announcement (given other players' announcements)

Auction 5 SA

Position 1				Position 2				Position 3			
i	v	$\hat{v}$	v^*	i	v	$\hat{v}$	v^*	i	v	$\hat{v}$	v^*
2	81	81	81	3	52	52	52	1	53	53	53
2	100	100	90	1	77	77	77	3	80	80	80
3	83	83	81	2	67	67	67	1	51	51	51
2	81	81	81	3	85	58	84	1	67	67	67
3	74	74	75	2	81	81	75	1	64	64	64
2	77	77	78	3	93	85	81	1	84	84	84
3	95	82	87	1	81	81	81	2	60	60	60
3	93	85	87	2	66	66	66	1	92	92	87
3	67	65	72	1	94	92	81	2	89	89	89
1	54	54	54	3	56	56	60	2	67	67	60
2	92	92	87	1	99	94	93	3	93	93	93
2	56	56	56	1	52	52	60	3	73	68	60
3	97	84	90	2	93	93	87	1	90	90	90
1	59	59	60	2	53	53	60	3	99	74	60
2	53	53	53	3	82	82	75	1	100	100	84
3	56	56	56	2	56	56	60	1	65	65	60
3	58	58	60	1	57	57	60	2	83	83	60
3	98	95	90	1	83	83	83	2	53	53	53
1	52	-	-	2	58	-	-	3	92	-	-
1	72	-	-	2	75	-	-	3	91	-	-

LEGEND

i: Player's number

v: Value

$\hat{v}$: Observed announcement

v^*: Equilibrium announcement (given other players' announcements)

In Round 18, Player 3 defected. He did not make any side payment.

Auction 6 BB

Player 1			Player 2			Player 3		
v	$\hat{b}$	b^*	v	$\hat{b}$	b^*	v	$\hat{b}$	b^*
93			64			88		
100	12	9	62	-12	2	82	6	8
99			60			92		
78			82			78		
73	10	5	53	-	0	88	10	6
91	9	9	54	-8	0	82	5	8
53			74			72		
69			57			89		
64			74			79		
92	10	8	79	-	7	59	-	0
65	-	3	81	7	4	58	-	0
88	8	10	57	-7	0	95	9	11
74	-	6	89	9	7	67	3	4
79	5	7	91	-11	11	93	9	11
58	-	0	86	8	6	71	-	5
70	3	5	53	-3	0	66	2	4
73			57			55		
69	-11	5	69	-8	5	95	9	6
95	9	9	82	8	8	63	1	2
81			52			71		

LEGEND

v: Value

$\hat{b} \geq 0$: Highest observed bid

$\hat{b} < 0$: Lowest observed ask

b^*: Equilibrium highest bid

-: Neither bid nor ask were made

Auction 7 LA

Player 1				*Player 2*				*Player 3*			
g	*v*	*s*	*g**	*g*	*v*	*s*	*g**	*g*	*v*	*s*	*g**
61	99	12	60	61	98	12	60	61	97	12	60
61	95	12	60	0	52	11	52	58	58	11	58
60	82	11	59	61	95	12	60	58	66	11	58
58	68	6	58	59	78	7	59	58	67	6	58
58	69	10	58	0	57	10	57	60	89	9	60
60	90	10	60	59	76	10	59	59	77	10	59
60	89	9	60	59	79	10	59	60	82	10	59
60	87	12	60	60	83	12	59	61	97	12	60
0	56	11	56	58	61	11	58	61	96	13	60
0	52	12	52	59	80	12	59	61	99	14	60
58	62	0	58	58	69	2	58	58	58	2	58
59	79	9	59	60	88	10	60	0	55	9	55
61	58	10	58	74	59	10	58	92	61	11	58
61	91	8	60	58	65	11	58	0	56	11	56
60	83	7	59	58	61	8	58	58	63	8	58
58	59	11	58	61	95	12	60	58	68	11	58
0	57	9	57	60	88	10	60	58	59	9	58
60	83	7	59	58	59	8	58	59	73	8	59
58	68	12	58	61	99	12	60	61	98	13	60
59	72	8	58	60	85	9	59	58	70	8	58

LEGEND

v: Value

g: Submitted bid (in the "legitimate" auction)

g^*: Equilibrium bid (given the side payment function)

s: Side Payment

Auction 8 SA

Position 1				Position 2				Position 3			
i	v	$\hat{v}$	v^*	i	v	$\hat{v}$	v^*	i	v	$\hat{v}$	v^*
1	59	59	60	2	75	75	72	3	91	91	78
1	97	97	90	2	93	93	93	3	54	54	54
2	89	89	84	3	60	60	60	1	71	71	71
1	65	65	69	3	51	51	51	2	80	80	66
1	88	88	84	3	71	71	71	2	92	92	90
1	98	98	90	2	82	82	82	3	59	59	59
3	78	78	78	2	69	63	69	1	66	66	66
2	60	60	63	1	60	60	63	3	60	60	60
1	91	91	87	2	71	71	71	3	71	71	71
1	82	82	81	2	71	71	71	3	69	69	69
3	88	88	84	1	60	60	60	2	53	53	53
1	58	58	60	3	78	78	72	2	88	84	81
1	62	62	66	2	51	51	51	3	76	76	63
1	89	89	84	2	69	69	69	3	89	89	90
1	86	86	84	3	85	85	87	2	87	87	87
1	71	71	72	3	87	87	78	2	92	88	90
2	78	78	78	1	62	62	62	3	82	82	81
3	53	53	53	2	58	58	60	1	65	65	60
2	56	56	56	1	72	72	69	3	95	95	75
1	99	99	90	2	90	90	90	3	90	90	90

LEGEND

i: Player's number

v: Value

$\hat{v}$: Observed announcement

v^*: Equilibrium announcement (given other players' announcements)

Auction 9 (1-12 and 20) SA

	Position 1				Position 2				Position 3		
i	v	$\hat{v}$	v^*	i	v	$\hat{v}$	v^*	i	v	$\hat{v}$	v^*
3	74	74	75	1	72	72	75	2	85	85	75
2	85	85	84	1	72	72	72	3	59	59	59
3	79	79	78	2	67	67	67	1	53	53	53
1	92	92	87	3	72	72	72	2	100	100	93
3	64	64	69	1	79	79	72	2	66	66	66
1	59	59	60	3	81	81	75	2	90	90	84
3	68	68	72	1	97	97	81	2	67	67	67
2	79	79	78	1	68	68	68	3	88	88	81
2	67	67	72	3	74	74	72	1	57	57	57
1	100	100	90	2	58	58	58	3	81	81	81
1	53	53	53	2	56	56	60	3	71	71	60
2	64	64	69	1	75	75	72	3	77	77	75
3	63	63	66	2	71	71	69	1	79	79	72

LEGEND
i: Player's number
v: Value
$\hat{v}$: Observed announcement
v^*: Equilibrium announcement (given other players' announcements)

Auction 9 (13-19) BB

Player 1			*Player 2*			*Player 3*		
v	$\hat{b}$	b^*	v	$\hat{b}$	b^*	v	$\hat{b}$	b^*
51			100			100		
93			70			70		
60	-	1	58	-	0	99	9	2
59	-	0	93	10	7	74	5	6
94	7	7	58	-7	0	74	5	6
74	5	7	74	6	7	96	7	8
82	7	8	67	3	4	100	9	9

LEGEND

v: Value

$\hat{b} \geq 0$: Highest observed bid

$\hat{b} < 0$: Lowest observed ask

b^*: Equilibrium highest bid

-: Neither bid nor ask were made

Auction 10 SA

	Position 1				Position 2				Position 3		
i	v	$\hat{v}$	v^*	i	v	$\hat{v}$	v^*	i	v	$\hat{v}$	v^*
2	88	88	84	1	98	98	89	3	61	61	61
2	78	78	78	3	64	64	64	1	78	78	78
2	58	58	60	3	76	76	72	1	76	76	76
3	58	58	60	1	82	82	75	2	71	71	71
2	97	97	90	1	72	72	72	3	75	75	75
3	61	61	63	1	99	99	84	2	66	66	66
3	61	61	63	1	77	77	72	2	57	57	57
1	57	57	57	3	77	77	72	2	65	65	65
3	76	76	78	2	85	85	78	1	69	69	69
2	89	89	84	1	86	86	86	3	83	83	83
1	62	62	66	2	63	63	66	3	71	71	66
3	65	65	69	1	78	78	72	2	58	58	58
3	82	82	81	2	53	53	53	1	76	76	76
3	79	79	78	2	76	76	76	1	76	76	76
3	82	82	81	2	88	88	84	1	55	55	55
2	76	76	78	3	96	96	81	1	64	64	64
3	68	68	72	2	90	90	78	1	71	71	71
3	63	63	66	2	78	78	72	1	76	76	76
3	76	76	78	1	91	91	78	2	53	53	53
3	83	83	81	1	79	79	79	2	97	97	84

LEGEND
i: Player's number
v: Value
$\hat{v}$: Observed announcement
v^*: Equilibrium announcement (given other players' announcements)

Auction 11 FO

Position 1				Position 2				Position 3			
i	v	$\hat{v}$	v^*	i	v	$\hat{v}$	v^*	i	v	$\hat{v}$	v^*
1	61	61	63	2	76	76	72	3	52	52	52
2	90	90	87	3	66	66	66	1	100	92	93
3	82	82	81	1	97	87	84	2	68	68	68
1	85	85	84	2	78	78	78	3	71	71	71
2	81	81	81	3	87	85	84	1	99	87	87
3	57	57	57	1	84	84	75	2	64	64	64
1	89	89	84	2	51	51	51	3	82	87	82
2	74	90	75	3	88	88	90	1	84	84	84
3	78	78	78	1	74	74	74	2	52	54	52
1	95	95	87	2	97	97	96	3	95	95	95
2	66	66	69	3	51	51	51	1	83	70	69
3	97	93	90	1	70	70	70	2	56	56	56
2	71	71	72	3	52	52	52	1	64	64	64
2	62	62	66	3	94	89	81	1	72	72	72
3	55	55	55	1	93	83	81	2	86	86	84
1	74	74	75	2	65	74	65	3	72	75	72
2	82	82	81	3	54	54	54	1	52	80	52
3	91	56	87	1	83	61	75	2	88	66	62
1	86	70	84	2	77	65	72	3	69	61	69
2	61	92	63	3	90	89	90	1	68	68	68

LEGEND

i: Player's number

v: Value

$\hat{v}$: Observed announcement

v^*: Equilibrium announcement (given other players' announcements)

Players stopped to cooperate after Round 15.

Auction 12 SA

Position 1				Position 2				Position 3			
i	*v*	*v̂*	*v**	*i*	*v*	*v̂*	*v**	*i*	*v*	*v̂*	*v**
1	60	60	63	2	70	70	69	3	92	92	72
1	56	56	56	2	88	88	78	3	79	79	79
1	86	86	84	2	92	92	87	3	98	76	93
2	51	51	51	3	82	82	75	1	68	68	68
3	54	54	54	1	70	70	69	2	82	82	72
2	63	63	66	3	87	87	78	1	94	91	90
1	58	58	60	3	74	74	72	2	75	75	75
3	61	61	63	2	78	78	72	1	73	73	73
2	90	90	87	3	80	80	80	1	55	55	55
2	85	85	84	1	58	58	58	3	57	57	57
2	74	74	75	3	74	74	75	1	60	60	60
2	96	96	90	3	61	61	61	1	97	97	96
2	75	75	75	3	51	51	51	1	86	80	78
2	91	91	87	3	80	80	80	1	52	52	52
1	70	70	72	2	79	79	72	3	64	64	64
2	86	86	84	1	51	51	51	3	72	72	72
3	93	88	87	2	55	55	55	1	82	82	82
2	61	61	63	3	81	78	75	1	94	92	81
2	91	91	87	1	86	86	86	3	51	51	51
1	72	72	75	3	82	75	75	2	54	54	54

LEGEND

i: Player's number

v: Value

v̂: Observed announcement

v*: Equilibrium announcement (given other players' announcements)

Auction 13 SA

Position 1				Position 2				Position 3			
i	v	$\hat{v}$	v^*	i	v	$\hat{v}$	v^*	i	v	$\hat{v}$	v^*
1	69	69	72	3	94	94	81	2	51	51	51
1	93	93	87	2	63	63	63	3	71	71	71
2	79	79	78	1	78	78	78	3	81	81	81
1	95	95	87	2	55	55	55	3	61	61	61
1	93	93	87	2	72	72	72	3	74	74	74
2	52	52	52	1	75	75	72	3	67	67	67
2	97	97	90	1	69	69	69	3	92	92	92
1	81	81	81	3	69	69	69	2	73	73	73
1	79	79	78	2	64	64	64	3	51	51	51
3	58	58	60	2	88	78	78	1	78	78	78
1	55	55	55	2	64	64	66	3	84	84	66
2	65	65	69	3	66	66	66	1	92	88	69
3	75	75	75	2	90	79	78	1	68	68	68
2	59	59	60	1	54	54	60	3	57	57	57
2	59	59	60	1	66	61	66	3	93	93	63
3	89	89	84	1	68	64	68	2	82	73	82
3	61	61	63	2	98	82	84	1	80	80	80
2	82	79	81	1	59	59	59	3	64	64	64
2	68	68	72	1	94	88	81	3	86	86	86
2	93	82	87	1	79	79	79	3	68	68	68

LEGEND
i: Player's number
v: Value
$\hat{v}$: Observed announcement
v^*: Equilibrium announcement (given other players' announcements)

Auction 14 SA

Position 1				*Position 2*				*Position 3*			
i	v	$\hat{v}$	v^*	i	v	$\hat{v}$	v^*	i	v	$\hat{v}$	v^*
2	69	69	72	3	69	69	72	1	82	82	72
2	81	81	81	3	100	100	84	1	93	93	93
2	92	92	87	1	65	65	65	3	78	78	78
2	100	100	90	3	59	59	59	1	53	53	53
1	80	80	81	3	70	70	70	2	55	55	55
2	100	100	90	1	59	59	59	3	84	84	84
1	66	66	69	2	52	52	52	3	52	52	52
1	60	60	63	2	91	91	78	3	54	54	54
2	94	94	87	1	80	80	80	3	65	65	65
2	68	68	72	1	59	59	59	3	53	53	53
1	59	59	60	3	94	94	81	2	78	78	78
1	78	78	78	2	70	70	70	3	54	54	54
1	98	98	90	2	51	51	51	3	65	65	65
1	51	51	51	3	89	89	78	2	87	87	87
2	91	91	87	1	79	79	79	3	63	63	63
1	99	99	90	2	77	77	77	3	86	86	86
1	63	63	66	2	86	86	78	3	99	99	87
1	66	66	69	2	85	85	75	3	85	85	85
3	99	99	90	1	67	67	67	2	69	69	69
1	62	62	66	3	71	71	69	2	52	52	52

LEGEND

i: Player's number

v: Value

$\hat{v}$: Observed announcement

v^*: Equilibrium announcement (given other players' announcements)

Auction 15 SA

Position 1				Position 2				Position 3			
i	v	$\hat{v}$	v^*	i	v	$\hat{v}$	v^*	i	v	$\hat{v}$	v^*
2	88	88	84	3	98	98	90	1	52	52	52
1	92	82	87	2	51	51	51	3	89	89	84
2	80	80	81	1	56	56	56	3	59	59	59
1	66	92	69	2	77	77	77	3	70	70	70
1	62	62	66	3	90	90	78	2	99	66	93
1	60	60	63	2	86	86	78	3	56	56	56
1	71	71	72	3	72	72	72	2	82	79	75
2	61	61	63	1	58	58	58	3	86	86	63
1	96	96	90	2	80	80	80	3	69	69	69
1	51	51	51	3	62	62	66	2	79	79	63
2	79	79	78	1	97	97	81	3	68	68	68
1	100	100	90	3	67	67	67	2	61	61	61
1	80	80	81	2	71	71	71	3	51	51	51
3	98	98	90	1	91	91	91	2	60	60	60
1	58	58	60	2	62	62	66	3	54	54	54
1	80	80	81	2	95	92	81	3	52	52	52
1	69	69	72	3	72	72	72	2	75	75	75
1	83	83	81	2	82	82	84	3	61	61	61
1	90	-	-	2	66	-	-	3	63	-	-
1	85	-	-	2	96	-	-	3	71	-	-

LEGEND

i: Player's number

v: Value

$\hat{v}$: Observed announcement

v^*: Equilibrium announcement (given other players' announcements)

Players stopped to cooperate after Round 18.

Auction 16 SA

Position 1				Position 2				Position 3			
i	v	$\hat{v}$	v^*	i	v	$\hat{v}$	v^*	i	v	$\hat{v}$	v^*
1	99	99	90	2	72	72	72	3	60	60	60
1	100	100	90	2	94	94	94	3	66	66	66
1	62	62	66	3	66	66	66	2	58	58	58
1	52	52	52	3	79	79	72	2	81	81	81
1	55	55	55	2	53	53	60	3	76	76	60
2	99	99	90	1	71	71	71	3	66	66	66
2	52	52	52	3	77	77	72	1	75	75	75
3	88	88	84	1	64	64	64	2	90	90	90
2	100	100	90	1	92	92	92	3	92	92	93
3	79	79	78	2	90	90	81	1	97	97	93
1	99	99	90	2	67	67	67	3	73	73	73
3	61	61	63	2	97	97	81	1	88	88	88
2	74	74	75	1	86	86	78	3	85	85	85
2	91	91	87	1	86	86	86	3	66	66	66
1	55	55	55	3	92	92	81	2	88	88	88
2	79	79	78	1	64	64	64	3	90	90	81
1	99	99	90	3	79	79	79	2	76	76	76
3	55	55	55	1	75	75	72	2	75	75	75
1	98	98	90	2	100	100	99	3	71	71	71
1	94	-	-	2	94	-	-	3	100	-	-

LEGEND
i: Player's number
v: Value
$\hat{v}$: Observed announcement
v^*: Equilibrium announcement (given other players' announcements)
Player 1 defected in Round 19. She bid 59 in the "legitimate" auction and did not make any side payment.

Appendix D

Non-cooperative Bidding

As we have already seen in Section 2.4.3, there are few rounds in which players bid non-cooperatively. In this appendix, we report the auctions and the rounds in which experimental subjects bid non-cooperatively. A legend, at the end of the table, explains the notation.

Auction	Round	v_1	b_1	b_1^*	v_2	b_2	b_2^*	v_3	b_3	b_3^*
5	19	90	70	77	66	58	61	63	70	60
5	20	85	72	73	96	90	81	71	70	64
6	1	93	83	79	64	61	60	88	70	75
6	3	99	86	83	60	58	58	92	88	78
6	4	78	70	69	82	76	72	78	76	69
6	17	73	65	66	57	57	57	55	55	55
6	20	81	71	71	52	50	52	71	69	64
11	17	52	57	52	82	61	72	54	57	54
11	18	83	62	72	88	68	75	91	70	77
11	19	86	71	74	77	72	68	69	60	63
11	20	68	68	63	61	58	59	90	79	77
15	19	52	0	52	58	58	58	92	75	78
15	20	72	68	65	75	68	67	91	75	77
16	20	94	64	79	94	75	79	100	85	83

LEGEND
v_i: Player i's value
b_i: Player i's submitted bid
b_i^*: Player i's risk neutral Nash equilibrium (RNNE) bid

Appendix E

Player 2's Equilibrium Strategy

In this appendix, we report Player 2's equilibrium strategy given that his value is v_2, Player 1 has announced $\hat{v}_1$, and Player 3 plays according to the strategy given in Proposition 1.

$v_2\backslash\hat{v}_1$	51	52	53	54	55	56	57	58	59	60
51	60	60	60	60	60	60	60	60	60	51
52	60	60	60	60	60	60	60	60	60	52
53	60	60	60	60	60	60	60	60	60	53
54	60	60	60	60	60	60	60	60	60	54
55	60	60	60	60	60	60	60	60	60	55
56	60	60	60	60	60	60	60	60	60	56
57	60	60	60	60	60	60	60	60	60	57
58	60	60	60	60	60	60	60	60	60	58
59	60	60	60	60	60	60	60	60	60	60
60	63	63	63	63	63	63	63	63	63	63
61	63	63	63	63	63	63	63	63	63	63
62	66	66	66	66	66	66	66	66	66	66
63	66	66	66	66	66	66	66	66	66	66
64	66	66	66	66	66	66	66	66	66	66
65	66	66	66	66	66	66	66	66	66	66
66	66	66	66	66	66	66	66	66	66	66
67	66	66	66	66	66	66	66	66	66	66
68	69	69	69	69	69	69	69	69	69	69
69	69	69	69	69	69	69	69	69	69	69
70	69	69	69	69	69	69	69	69	69	69
71	69	69	69	69	69	69	69	69	69	69
72	69	69	69	69	69	69	69	69	69	69
73	69	69	69	69	69	69	69	69	69	69
74	72	72	72	72	72	72	72	72	72	72
75	72	72	72	72	72	72	72	72	72	72

$v_2\backslash\hat{v}_1$	51	52	53	54	55	56	57	58	59	60
76	72	72	72	72	72	72	72	72	72	72
77	72	72	72	72	72	72	72	72	72	72
78	72	72	72	72	72	72	72	72	72	72
79	72	72	72	72	72	72	72	72	72	72
80	75	75	75	75	75	75	75	75	75	75
81	75	75	75	75	75	75	75	75	75	75
82	75	75	75	75	75	75	75	75	75	75
83	75	75	75	75	75	75	75	75	75	75
84	75	75	75	75	75	75	75	75	75	75
85	75	75	75	75	75	75	75	75	75	75
86	78	78	78	78	78	78	78	78	78	78
87	78	78	78	78	78	78	78	78	78	78
88	78	78	78	78	78	78	78	78	78	78
89	78	78	78	78	78	78	78	78	78	78
90	78	78	78	78	78	78	78	78	78	78
91	78	78	78	78	78	78	78	78	78	78
92	81	81	81	81	81	81	81	81	81	81
93	81	81	81	81	81	81	81	81	81	81
94	81	81	81	81	81	81	81	81	81	81
95	81	81	81	81	81	81	81	81	81	81
96	81	81	81	81	81	81	81	81	81	81
97	81	81	81	81	81	81	81	81	81	81
98	84	84	84	84	84	84	84	84	84	84
99	84	84	84	84	84	84	84	84	84	84
100	84	84	84	84	84	84	84	84	84	84

$v_2\backslash\hat{v}_1$	61	62	63	64	65	66	67	68	69	70
51	51	51	51	51	51	51	51	51	51	51
52	52	52	52	52	52	52	52	52	52	52
53	53	53	53	53	53	53	53	53	53	53
54	54	54	54	54	54	54	54	54	54	54
55	55	55	55	55	55	55	55	55	55	55
56	56	56	56	56	56	56	56	56	56	56
57	57	57	57	57	57	57	57	57	57	57
58	58	58	58	58	58	58	58	58	58	58
59	59	59	59	59	59	59	59	59	59	59
60	63	63	60	60	60	60	60	60	60	60
61	63	63	61	61	61	61	61	61	61	61
62	66	66	66	66	66	62	62	62	62	62
63	66	66	66	66	66	63	63	63	63	63
64	66	66	66	66	66	64	64	64	64	64
65	66	66	66	66	66	65	65	65	65	65
66	66	66	66	66	66	69	69	69	66	66
67	66	66	66	66	66	69	69	69	67	67
68	69	69	69	69	69	69	69	69	69	68
69	69	69	69	69	69	69	69	69	72	72
70	69	69	69	69	69	69	69	69	72	72
71	69	69	69	69	69	69	69	69	72	72
72	69	69	69	69	69	69	69	69	72	72
73	69	69	69	69	69	69	69	69	72	72
74	72	72	72	72	72	72	72	72	72	72
75	72	72	72	72	72	72	72	72	72	72

$v_2 \backslash \hat{v}_1$	61	62	63	64	65	66	67	68	69	70
76	72	72	72	72	72	72	72	72	72	72
77	72	72	72	72	72	72	72	72	72	72
78	72	72	72	72	72	72	72	72	72	72
79	72	72	72	72	72	72	72	72	72	72
80	75	75	75	75	75	75	75	75	75	75
81	75	75	75	75	75	75	75	75	75	75
82	75	75	75	75	75	75	75	75	75	75
83	75	75	75	75	75	75	75	75	75	75
84	75	75	75	75	75	75	75	75	75	75
85	75	75	75	75	75	75	75	75	75	75
86	78	78	78	78	78	78	78	78	78	78
87	78	78	78	78	78	78	78	78	78	78
88	78	78	78	78	78	78	78	78	78	78
89	78	78	78	78	78	78	78	78	78	78
90	78	78	78	78	78	78	78	78	78	78
91	78	78	78	78	78	78	78	78	78	78
92	81	81	81	81	81	81	81	81	81	81
93	81	81	81	81	81	81	81	81	81	81
94	81	81	81	81	81	81	81	81	81	81
95	81	81	81	81	81	81	81	81	81	81
96	81	81	81	81	81	81	81	81	81	81
97	81	81	81	81	81	81	81	81	81	81
98	84	84	84	84	84	84	84	84	84	84
99	84	84	84	84	84	84	84	84	84	84
100	84	84	84	84	84	84	84	84	84	84

$v_2\backslash\hat{v}_1$	71	72	73	74	75	76	77	78	79	80
51	51	51	51	51	51	51	51	51	51	51
52	52	52	52	52	52	52	52	52	52	52
53	53	53	53	53	53	53	53	53	53	53
54	54	54	54	54	54	54	54	54	54	54
55	55	55	55	55	55	55	55	55	55	55
56	56	56	56	56	56	56	56	56	56	56
57	57	57	57	57	57	57	57	57	57	57
58	58	58	58	58	58	58	58	58	58	58
59	59	59	59	59	59	59	59	59	59	59
60	60	60	60	60	60	60	60	60	60	60
61	61	61	61	61	61	61	61	61	61	61
62	62	62	62	62	62	62	62	62	62	62
63	63	63	63	63	63	63	63	63	63	63
64	64	64	64	64	64	64	64	64	64	64
65	65	65	65	65	65	65	65	65	65	65
66	66	66	66	66	66	66	66	66	66	66
67	67	67	67	67	67	67	67	67	67	67
68	68	68	68	68	68	68	68	68	68	68
69	72	69	69	69	69	69	69	69	69	69
70	72	70	70	70	70	70	70	70	70	70
71	72	72	71	71	71	71	71	71	71	71
72	72	75	75	75	72	72	72	72	72	72
73	72	75	75	75	73	73	73	73	73	73
74	72	75	75	75	75	74	74	74	74	74
75	72	75	75	75	75	75	75	75	75	75

$v_2\backslash\hat{v}_1$	71	72	73	74	75	76	77	78	79	80
76	72	75	75	75	78	78	78	76	76	76
77	72	75	75	75	78	78	78	78	77	77
78	72	75	75	75	78	78	78	78	78	78
79	72	75	75	75	78	78	78	81	81	81
80	75	75	75	75	78	78	78	81	81	81
81	75	75	75	75	78	78	78	81	81	81
82	75	75	75	75	78	78	78	81	81	81
83	75	75	75	75	78	78	78	81	81	81
84	75	75	75	75	78	78	78	81	81	81
85	75	75	75	75	78	78	78	81	81	81
86	78	78	78	78	78	78	78	81	81	81
87	78	78	78	78	78	78	78	81	81	81
88	78	78	78	78	78	78	78	81	81	81
89	78	78	78	78	78	78	78	81	81	81
90	78	78	78	78	78	78	78	81	81	81
91	78	78	78	78	78	78	78	81	81	81
92	81	81	81	81	81	81	81	81	81	81
93	81	81	81	81	81	81	81	81	81	81
94	81	81	81	81	81	81	81	81	81	81
95	81	81	81	81	81	81	81	81	81	81
96	81	81	81	81	81	81	81	81	81	81
97	81	81	81	81	81	81	81	81	81	81
98	84	84	84	84	84	84	84	84	84	84
99	84	84	84	84	84	84	84	84	84	84
100	84	84	84	84	84	84	84	84	84	84

$v_2 \backslash \hat{v}_1$	81	82	83	84	85	86	87	88	89	90
51	51	51	51	51	51	51	51	51	51	51
52	52	52	52	52	52	52	52	52	52	52
53	53	53	53	53	53	53	53	53	53	53
54	54	54	54	54	54	54	54	54	54	54
55	55	55	55	55	55	55	55	55	55	55
56	56	56	56	56	56	56	56	56	56	56
57	57	57	57	57	57	57	57	57	57	57
58	58	58	58	58	58	58	58	58	58	58
59	59	59	59	59	59	59	59	59	59	59
60	60	60	60	60	60	60	60	60	60	60
61	61	61	61	61	61	61	61	61	61	61
62	62	62	62	62	62	62	62	62	62	62
63	63	63	63	63	63	63	63	63	63	63
64	64	64	64	64	64	64	64	64	64	64
65	65	65	65	65	65	65	65	65	65	65
66	66	66	66	66	66	66	66	66	66	66
67	67	67	67	67	67	67	67	67	67	67
68	68	68	68	68	68	68	68	68	68	68
69	69	69	69	69	69	69	69	69	69	69
70	70	70	70	70	70	70	70	70	70	70
71	71	71	71	71	71	71	71	71	71	71
72	72	72	72	72	72	72	72	72	72	72
73	73	73	73	73	73	73	73	73	73	73
74	74	74	74	74	74	74	74	74	74	74
75	75	75	75	75	75	75	75	75	75	75

$v_2\backslash\hat{v}_1$	81	82	83	84	85	86	87	88	89	90
76	76	76	76	76	76	76	76	76	76	76
77	77	77	77	77	77	77	77	77	77	77
78	78	78	78	78	78	78	78	78	78	78
79	79	79	79	79	79	79	79	79	79	79
80	81	80	80	80	80	80	80	80	80	80
81	81	81	81	81	81	81	81	81	81	81
82	84	84	84	82	82	82	82	82	82	82
83	84	84	84	84	83	83	83	83	83	83
84	84	84	84	84	84	84	84	84	84	84
85	84	84	84	87	87	87	85	85	85	85
86	84	84	84	87	87	87	87	86	86	86
87	84	84	84	87	87	87	87	88	88	88
88	84	84	84	87	87	87	90	90	90	88
89	84	84	84	87	87	87	90	90	90	90
90	84	84	84	87	87	87	90	90	90	90
91	84	84	84	87	87	87	90	90	90	93
92	84	84	84	87	87	87	90	90	90	93
93	84	84	84	87	87	87	90	90	90	93
94	84	84	84	87	87	87	90	90	90	93
95	84	84	84	87	87	87	90	90	90	93
96	84	84	84	87	87	87	90	90	90	93
97	84	84	84	87	87	87	90	90	90	93
98	84	84	84	87	87	87	90	90	90	93
99	84	84	84	87	87	87	90	90	90	93
100	84	84	84	87	87	87	90	90	90	93

$v_2 \backslash \hat{v}_1$	91	92	93	94	95	96	97	98	99	100
51	51	51	51	51	51	51	51	51	51	51
52	52	52	52	52	52	52	52	52	52	52
53	53	53	53	53	53	53	53	53	53	53
54	54	54	54	54	54	54	54	54	54	54
55	55	55	55	55	55	55	55	55	55	55
56	56	56	56	56	56	56	56	56	56	56
57	57	57	57	57	57	57	57	57	57	57
58	58	58	58	58	58	58	58	58	58	58
59	59	59	59	59	59	59	59	59	59	59
60	60	60	60	60	60	60	60	60	60	60
61	61	61	61	61	61	61	61	61	61	61
62	62	62	62	62	62	62	62	62	62	62
63	63	63	63	63	63	63	63	63	63	63
64	64	64	64	64	64	64	64	64	64	64
65	65	65	65	65	65	65	65	65	65	65
66	66	66	66	66	66	66	66	66	66	66
67	67	67	67	67	67	67	67	67	67	67
68	68	68	68	68	68	68	68	68	68	68
69	69	69	69	69	69	69	69	69	69	69
70	70	70	70	70	70	70	70	70	70	70
71	71	71	71	71	71	71	71	71	71	71
72	72	72	72	72	72	72	72	72	72	72
73	73	73	73	73	73	73	73	73	73	73
74	74	74	74	74	74	74	74	74	74	74
75	75	75	75	75	75	75	75	75	75	75

$v_2 \backslash \hat{v}_1$	91	92	93	94	95	96	97	98	99	100
76	76	76	76	76	76	76	76	76	76	76
77	77	77	77	77	77	77	77	77	77	77
78	78	78	78	78	78	78	78	78	78	78
79	79	79	79	79	79	79	79	79	79	79
80	80	80	80	80	80	80	80	80	80	80
81	81	81	81	81	81	81	81	81	81	81
82	82	82	82	82	82	82	82	82	82	82
83	83	83	83	83	83	83	83	83	83	83
84	84	84	84	84	84	84	84	84	84	84
85	85	85	85	85	85	85	85	85	85	85
86	86	86	86	86	86	86	86	86	86	86
87	88	88	88	88	88	88	88	88	88	88
88	88	88	88	88	88	88	88	88	88	88
89	89	89	89	89	89	89	89	89	89	89
90	90	90	90	90	90	90	90	90	90	90
91	93	93	91	91	91	91	91	91	91	91
92	93	93	93	92	92	92	92	92	92	92
93	93	93	93	93	93	93	93	93	93	93
94	93	93	96	96	96	94	94	94	94	94
95	93	93	96	96	96	96	95	95	95	95
96	93	93	96	96	96	96	96	96	96	96
97	93	93	96	96	96	99	99	99	97	97
98	93	93	96	96	96	99	99	99	99	98
99	93	93	96	96	96	99	99	99	99	99
100	93	93	96	96	96	99	99	99	99	100

List of Figures

List of Tables